FERTIGATION TECHNOLOGIES FOR MICRO IRRIGATED CROPS

Performance, Requirements, and Efficiency

Innovations and Challenges in Micro Irrigation

FERTIGATION TECHNOLOGIES FOR MICRO IRRIGATED CROPS

Performance, Requirements, and Efficiency

Edited by
Megh R. Goyal, PhD, P.E.
Lala I. P. Ray, PhD

First edition published 2022

Apple Academic Press Inc.
1265 Goldenrod Circle, NE,
Palm Bay, FL 32905 USA
4164 Lakeshore Road, Burlington,
ON, L7L 1A4 Canada

CRC Press
6000 Broken Sound Parkway NW,
Suite 300, Boca Raton, FL 33487-2742 USA
2 Park Square, Milton Park,
Abingdon, Oxon, OX14 4RN UK

© 2022 Apple Academic Press, Inc.

Apple Academic Press exclusively co-publishes with CRC Press, an imprint of Taylor & Francis Group, LLC

Library and Archives Canada Cataloguing in Publication

Title: Fertigation technologies for micro irrigated crops : performance, requirements, and efficiency / edited by Megh R. Goyal, PhD, Lala I.P. Ray, PhD.

Names: Goyal, Megh R., editor. | Ray, Lala I. P., editor.

Series: Innovations and challenges in micro irrigation.

Description: First edition. | Series statement: Innovations and challenges in micro irrigation. | Includes bibliographical references and index.

Identifiers: Canadiana (print) 20200387030 | Canadiana (ebook) 20200387138 | ISBN 9781771889438 (hardcover) | ISBN 9781003084136 (ebook)

Subjects: LCSH: Microirrigation. | LCSH: Fertilizers.

Classification: LCC S619.T74 F47 2021 | DDC 631.5/87—dc23

Library of Congress Cataloging-in-Publication Data

Names: Goyal, Megh R., editor. | Ray, Lala I. P., editor.

Title: Fertigation technologies for micro irrigated crops : performance, requirements, and efficiency / Megh R. Goyal, Lala I. P. Ray.

Other titles: Innovations and challenges in micro irrigation.

Description: First edition. | Palm Bay, FL, USA : Apple Academic Press, 2021. | Series: Innovations and challenges in micro irrigation | Includes bibliographical references and index. | Summary: "Fertigation Technologies for Micro Irrigated Crops: Performance, Requirements, and Efficiency addresses the global water crisis by presenting new ways to use irrigation water judiciously through innovative fertigation management. It looks at the research and review works done throughout the world on micro irrigation and the techno-economic feasibility of various fertigation irrigation water management systems. Taking a multidisciplinary perspective, the chapters look at Using fertigation to increase the effectiveness of irrigation systems Crop performance evaluation of various crops under fertigation and irrigation methods Estimating levels of crop requirements Scheduling of fertigation and irrigation New fertigation equipment and technology Cost components of the various irrigation and fertigation systems This book will be a helpful resource for agricultural industry professionals, faculty, students, and others involved in irrigation and fertigation technology"-- Provided by publisher.

Identifiers: LCCN 2020050390 (print) | LCCN 2020050391 (ebook) | ISBN 9781771889438 (hardcover) | ISBN 9781003084136 (ebook)

Subjects: LCSH: Microirrigation. | Microirrigation--India. | Fertilizers.

Classification: LCC S619.T74 F478 2021 (print) | LCC S619.T74 (ebook) | DDC 631.5/87--dc23

LC record available at https://lccn.loc.gov/2020050390

LC ebook record available at https://lccn.loc.gov/2020050391

ISBN: 978-1-77188-943-8 (hbk)
ISBN: 978-1-77463-789-0 (pbk)
ISBN: 978-1-00308-413-6 (ebk)

OTHER BOOKS ON MICRO IRRIGATION TECHNOLOGY BY APPLE ACADEMIC PRESS, INC.

Book Series: Research Advances in Sustainable Micro Irrigation
Senior Editor-in-Chief: Megh R. Goyal, PhD, P.E.

Volume 1: Sustainable Micro Irrigation: Principles and Practices
Volume 2: Sustainable Practices in Surface and Subsurface Micro Irrigation
Volume 3: Sustainable Micro Irrigation Management for Trees and Vines
Volume 4: Management, Performance, and Applications of Micro Irrigation Systems
Volume 5: Applications of Furrow and Micro Irrigation in Arid and Semi-Arid
        Regions
Volume 6: Best Management Practices for Drip Irrigated Crops
Volume 7: Closed Circuit Micro Irrigation Design: Theory and Applications
Volume 8: Wastewater Management for Irrigation: Principles and Practices
Volume 9: Water and Fertigation Management in Micro Irrigation
Volume 10: Innovation in Micro Irrigation Technology

Book Series: Innovations and Challenges in Micro Irrigation
Senior Editor-in-Chief: Megh R. Goyal, PhD, P.E.

- Engineering Interventions in Sustainable Trickle Irrigation:
  Water Requirements, Uniformity, Fertigation, and Crop Performance
- Fertigation Technologies for Micro Irrigated Crops:
  Performance, Requirements, and Efficiency
- Management Strategies for Water Use Efficiency and Micro Irrigated Crops:
  Principles, Practices, and Performance
- Micro Irrigation Engineering for Horticultural Crops: Policy Options,
  Scheduling and Design
- Micro Irrigation Management: Technological Advances and Their Applications
- Micro Irrigation Scheduling and Practices
- Performance Evaluation of Micro Irrigation Management:
  Principles and Practices
- Potential of Solar Energy and Emerging Technologies in Sustainable
  Micro Irrigation
- Principles and Management of Clogging in Micro Irrigation
- Sustainable Micro Irrigation Design Systems for Agricultural Crops:
  Methods and Practices

ABOUT THE SENIOR EDITOR-IN-CHIEF

Megh R. Goyal, PhD

Retired Professor in Agricultural and Biomedical Engineering, University of Puerto Rico, Mayaguez Campus; Senior Acquisitions Editor, Biomedical Engineering and Agricultural Science, Apple Academic Press, Inc.

Megh R. Goyal, PhD, PE, is a Retired Professor in Agricultural and Biomedical Engineering from the General Engineering Department in the College of Engineering at the University of Puerto Rico–Mayaguez Campus; and Senior Acquisitions Editor and Senior Technical Editor-in-Chief in Agriculture and Biomedical Engineering for Apple Academic Press, Inc. He has worked as a Soil Conservation Inspector and as a Research Assistant at Haryana Agricultural University and Ohio State University.

During his professional career of 49 years, Dr. Goyal has received many prestigious awards and honors. He was the first agricultural engineer to receive the professional license in Agricultural Engineering in 1986 from the College of Engineers and Surveyors of Puerto Rico. In 2005, he was proclaimed as "Father of Irrigation Engineering in Puerto Rico for the Twentieth Century" by the American Society of Agricultural and Biological Engineers (ASABE), Puerto Rico Section, for his pioneering work on micro irrigation, evapotranspiration, agroclimatology, and soil and water engineering. The Water Technology Centre of Tamil Nadu Agricultural University in Coimbatore, India, recognized Dr. Goyal as one of the experts "who rendered meritorious service for the development of micro irrigation sector in India" by bestowing the Award of Outstanding Contribution in Micro Irrigation. This award was presented to Dr. Goyal during the inaugural session of the National Congress on "New Challenges and Advances in Sustainable Micro Irrigation" held at Tamil Nadu Agricultural University. Dr. Goyal received the Netafim Award for Advancements in Microirrigation: 2018 from the American Society of Agricultural Engineers at the ASABE International Meeting in August 2018.

A prolific author and editor, he has written more than 200 journal articles and textbooks and has edited over 85 books. He is the editor of three book series published by Apple Academic Press: Innovations in Agricultural & Biological Engineering, Innovations and Challenges in Micro Irrigation, and Research Advances in Sustainable Micro Irrigation. He is also instrumental in the development of the new book series Innovations in Plant Science for Better Health: From Soil to Fork.

Dr. Goyal received his BSc degree in engineering from Punjab Agricultural University, Ludhiana, India; his MSc and PhD degrees from Ohio State University, Columbus; and his Master of Divinity degree from Puerto Rico Evangelical Seminary, Hato Rey, Puerto Rico, USA.

ABOUT CO-EDITOR

Lala I. P. Ray, PhD

Lala I. P. Ray, PhD, is currently working as Associate Professor in Soil and Water Engineering, School of Natural Resource Management, College of Postgraduate Studies in Agricultural Sciences, Central Agricultural University-Imphal, Umiam, Meghalaya, India.

His BTech degree in Agricultural Engineering was from Odisha University of Agricultural University, Bhubaneswar, Odisha, India. With ICAR-JRF fellowship, he received his master in Soil and Water Conservation at Tamil Nadu Agricultural University, Coimbatore, Tamil Nadu. His PhD degree on Integrated Water Management was from the Indian Institute of Technology, Kharagpur (IIT Kharagpur).

He has expertise in the principles and practices of water management, soil conservation, and water quality aspects. He has worked at the ICAR-Indian Institute of Water Management in an ad hoc project sponsored by the Ministry of Water Resources, Government of India, as Research Associate for two years. He also has worked as an Agriculture Officer in a Livelihood project funded by DFID under United Nations, Odisha Cell. At present, he is actively involved in teaching, research, and extension activities at Central Agricultural University-Imphal.

He has worked extensively on a gravity-fed drip irrigation system for the hilly states of North Eastern India through a Govt. of India (Department of Science and Technology) funded project. He prepared contingency crop planning for seven districts of Meghalaya by a meticulous weather parameter analysis.

He is recipient of various awards, such as National Fellowship by Government of India; ICAR-JRF fellowship for his master degree; best student award from Tamil Nadu Agricultural University; Dr. Ismail Serageldin Award for best student in soil and water conservation engineering; best paper award by the Indian Association of Hill Farming (IAHF), Meghalaya; and Israel Government Fellowship for pursuing his higher Research program at Elat, Israel.

Readers may contact him at: <lalaipray@rediffmail.com>

CONTENTS

CONTRIBUTORS

Rajan Aggarwal
Senior Research Engineer, Department of Soil and Water Engineering, College of Agricultural Engineering and Technology, Punjab Agricultural University (PAU), Ludhiana 141004, Punjab, India; E-mail: rajaaggarwal1@gmail.com

Mukhtar Ahmed
Assistant Professor, Department of Agronomy, Pir Mehr Ali Shah (PMAS) Arid Agriculture University, Rawalpindi 46300, Pakistan; Mobile: +92-3335959893; E-mail: ahmadmukhtar@uaar.edu.pk

Shakeel Ahmad
Post Doctorate, Department of Agronomy, Bahauddin Zakariya University, Multan 60800, Pakistan; Mobile: +92-3007318911; E-mail: shakeelahmad@bzu.edu.pk

Pankaj Barua
Director, North Eastern Regional Institute of Water and Land Management (NERIWALAM), Assam Agricultural University (AAU), P.O. Kaliabhomora, Dolabari 784027, Tezpur, Assam, India; Mobile: +91-94350-83111; E-mail: pankaj_barua@hotmail.com

Yatnesh Bisen
Senior Research Fellow, M2M Project, Department of Soil and Water Engineering, SV College of Agricultural Engineering and Technology & Research Station, Faculty of Agricultural Engineering, Indira Gandhi Krishi Vishwavidyalaya, Raipur 492012, Chhattisgarh, India; Mobile: +91-9575140099; E-mail: yatnesh12@gmail.com

R. Chelleng
M. Tech. Student, Department of Agricultural Engineering, Assam University, Silchar 788011, Assam, India; Mobile: +917086831253; E-mail: racktack289@gmail.com

A. Dalai
PhD Research Scholar, Department of Soil and Water Engineering, SVCAET & RS, IGKV, Raipur 492012, Chhattisgarh, India; Mobile: +91-9556649550; E-mail: dalaiabinash@gmail.com

Benukantha Dash
M. Tech, Scientist, ICAR- National Bureau of Soil Survey and Land Use Planning, Nagpur 440033, India; Mobile: +91-94374-02827; E-mail: benukantha@yahoo.co.in

P. Debnath
Professor, College of Horticulture and Forestry, Central Agricultural University (CAU)-Imphal, Pasighat 791102, Arunachal Pradesh, India; Mobile: +91-9402477047; E-mail: kanupran@yahoo.co.in

Shah Fahad
Post Doctorate, Department of Agriculture, University of Swabi, Khyber Pakhtunkhwa, Pakistan; Mobile: +92-3009304952; E-mail: shah_fahad80@yahoo.com

Suchi Gangwar
Research Associate, Irrigation and Drainage Department, ICAR - Central Institute of Agricultural Engineering (CIAE), Precision Farming Development Centre (PFDC), Nabi Bagh, Berasia Road, Bhopal 462038, Madhya Pradesh, India; Mobile: +91-94075-22821; E-mail: singh.suchi40@gmail.com

N. V. Gowtham Deekshithulu
PhD Research Scholar, Department of Soil and Water Engineering, Dr NTR College of Agricultural
Engineering, Bapatla 522101, Andhra Pradesh, India; E-mail: gowtham.deekshithulu@gmail.com

Megh R. Goyal
Retired Faculty in Agricultural and Biomedical Engineering from College of Engineering at University
of Puerto Rico–Mayaguez Campus; and Senior Technical Editor-in-Chief in Agricultural and
Biomedical Engineering for Apple Academic Press Inc.; PO Box 86, Rincon–PR–006770086, USA;
E-mail: goyalmegh@gmail.com

Fayyaz-ul-Hassan
Post Doctorate, Pir Mehr Ali Shah (PMAS) Arid Agriculture University, Rawalpindi 46300, Pakistan;
Mobile: +92-3009514597; E-mail: drsahi63@gmail.com

Santosh Kumar
Subject Matter Specialist (Horticulture), Krishi Vigyan Kendra (KVK) at Aizawl, Central Agricultural
University (CAU), Imphal, C.V.Sc. & A.H., Selesih 796014, Mizoram, India; Mobile: +91-8287629636;
E-mail: santosh@gmail.com

Avinash Kumar
Assistant Professor, Department of Agricultural Engineering, Assam University, Silchar 788011, Assam,
India; Mobile: +919435779233; E-mail: avinashiitkgp86@gmail.com

Dhiraj Khalkho
Scientist, AICRP on Irrigation Water Management, Department of Soil and Water Engineering,
SV College of Agricultural Engineering and Technology & Research Station, Faculty of Agricultural
Engineering, Indira Gandhi Krishi Vishwavidyalaya, Raipur 492012, Chhattisgarh, India;
Mobile: +91-9826534139; E-mail: dkhalkho@rediffmail.com

E.K Kurien
Professor and Head, Agronomic Research Station, Chalakudy 691301, Kerala, India;
Mobile: +91-9447614627; E-mail: ek.kurien@kau.in

M. Manikandan
Assistant Professor, Department of Irrigation and Drainage Engineering, Agricultural Engineering College
and Research Institute (AEC & RI), Tamil Nadu Agricultural University (TNAU), Kumulur 621712,
Tamil Nadu, India; Mobile: +91-9486620044; E-mail: muthiahmanikandan29@gmail.com

Jotish Nongthombam
Subject Matter Specialist (Agricultural Engineering), Krishi Vigyan Kendra (KVK) at Aizawl,
Central Agricultural University (CAU), Imphal, C.V.Sc. & A.H., Selesih 796014, Mizoram, India;
Mobile: +91-82570-38313; E-mail: jnongthombam@gmail.com.

E. K. Mathew
Professor (Retired), Kelappaji College of Agricultural Engineering and Technology (KCAET),
Kerala Agricultural University (KAU), P.O. Malappuram, Thavanur 679573, Kerala, India;
Mobile: +91- 9447233692; E-mail: ekmathew@yahoo.com

R. K. Naik
Scientist, AICRP on FIM, Department of Farm Machinery and Power Engineering, SV College of
Agricultural Engineering and Technology & Research Station, Faculty of Agricultural Engineering,
Indira Gandhi Krishi Vishwavidyalaya, Raipur 492012, Chhattisgarh, India;
Mobile: +91-7470986744; E-mail: rknaik1@rediffmail.com

Narinder K. Narda
Professor (Retired), Department of Soil and Water Engineering, College of Agricultural Engineering
and Technology, Punjab Agricultural University, Ludhiana–141004, Punjab, India;
Mobile: +91-9996457556; E-mail: narinder697@gmail.com

Nadiya Nesthad
Assistant Professor, Kochu Koikkal Puthen House, Paravoor, Kollam 691301, Kerala, India;
Mobile: +91-9496020544; E-mail: nadiyanesthad@gmail.com

Vyas Pandey
Emeritus Scientist (ICAR) and Former Professor and Head, Department of Agricultural Meteorology,
Anand Agricultural University (AAU), Anand 388110, Gujarat, India; Mobile: +91-9879912357;
E-mail: vpandey@aau.in; pandey04@yahoo.com; vyask.pandey@gmail.com;

Balram Panigrahi
Professor and Head Department of Soil and Water Conservation Engineering, College of Agricultural
Engineering and Technology, Odisha University of Agriculture and Technology, Bhubaneswar 751003,
Odisha, India; Mobile: +91-94378-82699; E-mail: kajal_bp@yahoo.co.in

S. K. Patil
Vice Chancellor, Indira Gandhi Krishi Vishwavidyalaya, Raipur 492012, Chhattisgarh, India;
Mobile: +91-7712443419; E-mail: spatil_igau@yahoo.com

S. K. Pattanaaik
Associate Professor, College of Horticulture and Forestry, Central Agricultural University
(CAU)- Imphal, Pasighat 791102, Arunachal Pradesh, India; Mobile: +91-9436630596;
E-mail: saroj_swce@rediffmail.com

P. C. Pradhan
Junior Scientist, Department of Soil and Water Conservation Engineering, Precision Farming
Development Centre (PFDC), Odisha University of Agriculture and Technology (OUAT),
Bhubaneswar 751003, Odisha, India; Mobile: +91-9438302925; E-mail: pcpradhan@gmail.com

K. V. Ramana Rao
Principal Scientist, Irrigation and Drainage Department, ICAR - Central Institute of Agricultural
Engineering (CIAE), Precision Farming Development Centre (PFDC), Nabi Bagh, Berasia Road,
Bhopal 462038, MP, India; Mobile: +91-94250-23739, E-mail: kvramanarao1970@gmail.com

G. Ravi Babu
Professor and Head, Department of Irrigation and Drainage Engineering, Dr NTR College of
Agricultural Engineering, Bapatla 522101, Andhra Pradesh, India; Mobile: +91 9848572321;
E-mail: ravibg2004@yahoo.com,

Lala I. P. Ray
Associate Professor (Soil and Water Engineering), College of Postgraduate Studies in Agricultural
Sciences, (Central Agricultural University-Imphal), Umiam-793103, Meghalaya, India;
Mobile: +91-94363-36021; E-mail: lalaipray@rediffmail.com

Raja Gopala Reddy
Senior Research Fellow, Department of Agricultural and Food Engineering, Indian Institute of
Technology (IIT), Kharagpur 721301, West Bengal, India; Mobile: 91- 9493413168;
E-mail: rajbckv@gmail.com

Narayan Sahoo
Professor, College of Agricultural Engineering and Technology, Odisha University of Agriculture and
Technology, Bhubaneswar 751003, Odisha, India; Mobile: +91-94371-91308;
E-mail: narayan_swce@yahoo.co.in

Manoj P. Samuel
Principal Scientist, ICAR-Central Institute of Fisheries Technology, Kochi 682029, Kerala, India;
Mobile: +91-9177943425; E-mail: manojpsamuel@gmail.com

Santosh D. T.
PhD Research Scholar, Department of Agricultural and Food Engineering,
Indian Institute of Technology (IIT), Kharagpur 721301, West Bengal, India;
Mobile: 91-9735267433; E-mail: dtsantosh@gmail.com

Sudipto Sarkar
Associate Professor, Department of Agricultural Engineering, Assam University, Silchar 788011,
Assam, India; Mobile: +919435179371; E-mail: sudiptoiit@gmail.com

R. K. Singh
Principal Scientist, ICAR-Central Arid Zone Research Institute, Jodhpur 342003, Rajasthan, India;
Mobile: +91-7726953529; E-mail: rksinghiinrg@gmail.com

A. Suresh
Principal Scientist, ICAR-Central Institute of Fisheries Technology, Kochi 682029, Kerala, India;
Mobile: +91-7838963081; E-mail: sureshcswri@gmail.com

Arati Sethi
M. Tech. Student, College of Agricultural Engineering and Technology, Odisha University of
Agriculture and Technology, Bhubaneswar 751003, Odisha, India; Mobile:+91-70083-03079;
E-mail: arati.sethi89@gmail.com

Laxmi Narayan Sethi
Professor, Department of Agricultural Engineering, Assam University, Silchar 788011, Assam, India;
Mobile: +919401847943; E-mail: lnsethi06@gmail.com

D. S. Thakur
Dean, College of Horticulture, Indira Gandhi Krishi Vishwavidyalaya, Jagdalpur 494005,
Chhattisgarh, India; Mobile: +91-9424270404; E-mail: dstigkv@rediffmail.com

G. Thiyagarajan
Scientist-cum-Assistant Professor, Water Technology Centre, Tamil Nadu Agricultural University
(TNAU), Coimbatore 641003, Tamil Nadu, India; Mobile: +91-948657065; E-mail: thiyagu@tnau.ac.in.

Kamlesh N. Tiwari
Professor, Department of Agricultural and Food Engineering, Indian Institute of Technology (IIT),
Kharagpur 721301, West Bengal, India; Mobile: 91-3222283150; E-mail: kamlesh_iitkgp@yahoo.co.in

Anu Varghese
Assistant Professor, Kelappaji College of Agricultural Engineering and Technology (KCAET),
Kerala Agricultural University (KAU), P.O. Malappuram, Thavanur 679573, Kerala, India;
Mobile: +91- 9495174921; E-mail: anu.varughese@kau.in

Ajay Kumar Vashisht
Associate Professor, Department of Irrigation and Drainage Engineering,
College of Agricultural Engineering and Postharvest Technology,
Central Agricultural University (CAU) Imphal, Ranipool–737135, Sikkim, India;
Mobile: +91-8016269685; E-mail: akvashisht74@yahoo.com

Anamika Yadav
PhD Student, Department of Agricultural Engineering, Assam University, Silchar 788011, Assam, India;
Mobile: +917987178532; E-mail: anamika.iit26@gmail.com

ABBREVIATIONS

σ	Stefan-Boltzmann constant ($4.903\ 10^{-9}\,MJ\ m^{-2}\,day^{-1}$)
$(NH4)_2SO_4$	ammonium sulfate
a	constant used in Thornthwaite method
AICRPIWM	All India Coordinated Research Project on Irrigation Water Management
Al	aluminum
AOAC	Association of Official Agricultural Chemists
APSIM	Agricultural Production Systems Simulator
a_s	fraction of extraterrestrial radiation reaching the earth on an overcast day
$a_s + b_s$	fraction of extraterrestrial radiation reaching the earth on a clear day
B	boron
$B(OH)_3$	boric acid
$B(OH)^{4-}$	borate
B:C	benefit-cost
BCR	benefit to cost ratio
BD	bulk density
BIS	Bureau of Indian Standards
BSS	bright sunshine hour (hours)
c	constant used in FAO radiation method for mean relative humidity and wind speed
C.V.Sc. & A.H.	College of Veterinary Science & Animal Husbandry
Ca	calcium
$Ca\ (NO_3)^2$	calcium nitrate
$Ca\ (NO_3)^2$ $\cdot NH_4NH_3 \cdot 10H_2O$	calcium ammonium nitrate
CaCl	calcium chloride
$CaCO_3$	calcium carbonate
CAU	Central Agricultural University
C_H	humidity coefficient
CIAE	Central Institute of Agricultural Engineering
Cl	chloride
$CO(NH_2)_2$	urea

CP	crop production (ton year^{-1}) of the country
C_p	specific heat at constant pressure (1.013×10^{-3} MJ kg^{-1} °C^{-1})
C_s	solar radiation coefficient.
CSS	Centrally Sponsored Scheme
C_T	temperature coefficient
Cu	copper
CU	coefficient of uniformity
cum	cubic meter
CV	coefficient of variation
Cv	coefficient of variation
C_w	wind velocity coefficient
CWR	crop water requirement
CWU	crop water use
CY	crop yield
DAS	day after sowing
df	degree of freedom
d_r	inverse relative distance Earth-Sun (radian)
dSm^{-1}	Deci Siemen per meter
DSSAT	Decision Support System for Agrotechnology Transfer
DTMF	Dual Tone Multiple Frequency
$e_{(Tmax)}$	saturation vapor pressure at daily maximum temperature (kPa)
$e_{(Tmin)}$	saturation vapor pressure at daily minimum temperature (kPa)
e_a	actual vapor pressure (kPa)
E_a	isothermal evaporation rate used in Penman method (kg m^{-2} s^{-1})
EC	electrical conductivity (mS)
EDDHA	ethylene-diamine di ortho-hydroxyphenylacetic acid
EDTA	ethylene-diamine-tetraacetic acid
E_{pan}	pan evaporation (mm or mm day^{-1})
$e_s - e_a$	saturation vapor pressure deficit (kPa)
e_s	saturation vapor pressure (kPa)
ET	evapotranspiration
ETc	crop evapotranspiration (mm day^{-1})
ETo	reference evapotranspiration (mm day^{-1})
FAO	Food and Agriculture Organization
FAO-STAT	Food and Agriculture Organization Corporate Statistical Database

FC	field capacity
Fe	iron
Fe^{2+}	ferrous ion
Fe^{3+}	ferric ion
FPP	flexible polypropylene
FPPR	reinforced flexible polypropylene
FUE	fertilizer use efficiency
G	soil heat flux density ($MJ\ m^{-2}\ day^{-1}$)
GDS	gravity type drip irrigation system
GGGI	Global Green Growth Institute
GHG	green house gas
GIS	Geographical Information System
GoI	Government of India
GPS	Geographical Position System
G_{sc}	solar constant ($0.0820\ MJ\ m^{-2}\ min^{-1}$)
GSM	Global System for Mobile Communication
GSWD	gross specific water demand
H	height of the measurement above the ground surface (m)
h	mean plant height
H^+	hydrogen ion
H_2PO^{4-}	dihydrogen phosphate
H_3PO_4	phosphoric acid
HCO_3	bicarbonate
HDPE	high density polyethylene
HMNE	Horticulture Mission for North East India
HPO_4^{2-}	hydrogen phosphate
I	heat index
I	irrigation
ICAR	Indian Council of Agricultural Research
IE	irrigation efficiency
I_E	effective irrigation
IFS	Integrated Farming System
I_G	gross irrigation
IINRG	Indian Institute of Natural Resins and Gums
ILRI	Indian Lac Research Institute
I_N	net irrigation
IST	Indian Standard Time
IWMI	International Water management Institute
IWR	irrigation water requirement

IWUE	irrigation water use efficiency
IWUE	irrigation water use efficiency (t/ha/mm)
J	Julian day/Number of the day in the year
K	potassium
k	monthly consumptive use coefficient used in Blaney-Criddle method
K_2O	potash
K_2SO_4	potassium sulfate
Kc	crop coefficient
KCAET	Kelappaji College of Agricultural Engineering and Technology
Kc_{end}	crop coefficient at the end of growing period
KCl	potassium chloride
Kc_{mid}	crop coefficient at mid-season growth stage of crop
kg ha^{-1}	kilogram per hectare
kg/ha	kilogram per hectare
KH_2PO_4	mono potassium phosphate (MKP)
KNO3	potassium nitrate
K_p	pan coefficient
KVK	Krishi Vigyan Kendra
LAI	leaf area index
LCD	liquid crystal display
LDPE	low density polyethylene
LED	light emitting diode
LLDPE	linear low density polyethylene
LMT	local mean time
lph	liter per hour
m	number of days in a month
MAD	management allowable depletion
MCU	microcontroller unit
M_f	monthly latitude dependent factor in Hargreaves method
Mg	magnesium
$Mg(NO_3)^2$	magnesium nitrate
MI	micro irrigation
MJ	mega - Joules
MKP	mono potassium phosphate
mmol	milli mole
Mmol/l	milli mol per liter
Mn^{2+}	manganese ion

MOP	muriate of potash
MSL	mean sea level
MWH	micro water harvesting
N	nitrogen
n	actual duration of sunshine hours (hours)
N	daylight hours/maximum possible sunshine hours (hours)
N_2	nitrogen
N_2O	nitrous oxide
Na	sodium
NABARD	National Bank for Agriculture and Rural Development
NaCl	sodium chloride
NEH	North East Hilly
NERIWALAM	North Eastern Regional Institute of Water and Land Management
NH_3	ammonia
NH^{4+}	ammonium ion
NH_4CO_3	ammonium carbonate
$NH_4H_2PO_4$	mono ammonium phosphate (MAP)
NH_4NO_3	ammonium nitrate
NH4OH	ammonium hydroxide
Ni	nickel
NIH	National Institute of Hydrology
NMMI	National Mission on Micro Irrigation
NMSA	National Mission on Sustainable Agriculture
NO	nitric oxide
NO^{3-}	nitrate ion
NPK	Nitrogen (N), Phosphorous (P) and Potassium (K)
OC	organic carbon
OC	organic content
Op-amp	Operational amplifier
P	phosphorus
P	atmospheric pressure (kPa)
p	mean monthly percentage of annual day time hours
PAU	Punjab Agricultural University
PAW	plant available water
P_d	deep percolation
PDC	Plasticulture Development Center
PE	pan evaporation
PET	potential evapotranspiration

PFDC	Precision Farming Development Centre
pH	H^+ ion concentration (a measure of acidity or alkalinity)
pH	potential of hydrogen
P-M	Penman Monteith
PMKSY	Prime Minister's *Krishi Sinchayee Yojana*
P-T	Priestley Taylor
PTEF	poly tetra fluorethylene
PVC	polyethylene or poly vinyl chloride
PWP	permanent wilting point
PWR	plant water requirement
q	Quintal, = 100 kg
qtl	Quintal, = 100 kg
R	rainfall
R_a	extraterrestrial radiation (MJ m^{-2} day^{-1})
RAW	readily available water
RDF	recommended dose of fertilizer
RDN	recommended dose of Nitrogen
R_E	effective Rainfall
RF	radio frequency
RFD	recommended fertilizer dose
RH	relative humidity (%)
RH_{max}	morning relative humidity/Relative humidity at 0738 hours
RH_{min}	afternoon relative humidity/Relative humidity at 1438 hours
R_n	net radiation (MJ m^{-2} day^{-1})
R_N	net Rainfall
R_{nl}	net outgoing long wave radiation (MJ m^{-2} day^{-1})
R_{ns}	net solar or shortwave radiation (MJ m^{-2} day^{-1})
RO	surface runoff
RRWH	rooftop rain water harvesting
R_s	solar or shortwave radiation (MJ m^{-2} day^{-1})
R_{so}	clear-sky solar radiation (MJ m^{-2} day^{-1})
S	storage
SD	standard deviation
SHU	Scoville Heat Unit
SMS	short message service
SMW	standard meteorological week
SO_4^{2-}	sulfate
SSDI	sub-surface drip irrigation
SSI	Sustainable Sugarcane Initiative

SWC	soil water content
SWD	specific water demand (m^3 ton^{-1}) of a crop in the country
T	mean air temperature (daily, mean monthly) (°C)
T_d	difference between maximum and minimum daily air temperature (°C)
T_{max}	maximum daily air temperature (°C),
T_{min}	minimum daily air temperature (°C),
TNAU	Tamil Nadu Agricultural University
tons/ha	tons per hectare
TSS	total soluble solids
U	up-flux from shallow groundwater table
u_2	wind speed 2 m above the ground surface ($m\ s^{-1}$)
USDA	United States Department of Agriculture
u_z	measured wind speed at z meter above the ground surface ($km\ s^{-1}$)
VPD	vapor pressure deficit (kPa)
VWR	virtual water requirement (m^3 $year^{-1}$) of the country
W	constant used in FAO radiation method for temperature and altitude
WAS	weeks after sowing
WF	water foot printing
WR	water requirement
WSN	wireless sensor network
WUE	water use efficiency
Z	elevation above sea level/Altitude (m)
Zn	zinc
α	albedo or canopy reflection coefficient
Δ	slope saturation vapor pressure curve ($kPa°C^{-1}$)
δ	solar declination (radian)
ε	ratio molecular weight of water vapor/dry air (0.622)
λ	latent heat of vaporization ($MJ\ kg^{-1}$)
Υ	psychrometric constant ($kPa°C^{-1}$)
φ	latitude (radian)
ω_s	sunset hour angle (radian)

PREFACE

The water crisis is a global concern for all of us who are working in the agricultural sector, and this issue rings the alarm bell. How to use irrigation water judiciously is a million dollar question. Under the scenario of climate change, how to address this by irrigation water management and innovative ways of field water management means is compiled in this edited book.

At this present juncture, this book is edited to scatter more ideas and knowledge on innovative ways of irrigation water management and its techno-economic feasibility. Recent advancements on field water management, drip fertigation, automation, and micro irrigation practices need attention by the irrigation community. This book will be a helpful guide to undergraduate, post-graduate, and personnel from industries.

This edited book is an encapsulation of compiled research and review works done throughout the world on micro irrigation and water management. This book has been prepared under four different sections consisting of seventeen chapters, e.g.: (1) Crop performance under micro irrigation systems; (2) Irrigation requirement of drip irrigated crops; (3) Automation and fertigation technologies in micro irrigation; and (4) Enhancement of irrigation efficiency. Forty-one researchers have contributed on various issues from crop performance and crop water requirements with and without micro irrigation system. Advanced research on automation with micro irrigation system has been discussed lucidly. The cost components of the various irrigation parts have been incorporated. The various policy issues related to micro irrigation system in particular have been discussed.

At the 49th annual meeting of the Indian Society of Agricultural Engineers at Punjab Agricultural University on February 22–25 of 2015, a group of ABEs convinced me that there is a dire need to publish book volumes on the focus areas of agricultural and biological engineering (ABE), including irrigation. This is how the idea was born on new book series titled *Innovations and Challenges in Micro Irrigation*.

The contributions by the cooperating authors to this book volume have been most valuable in the compilation. Their names are mentioned in each chapter and in the list of contributors. This book would not have been written without the valuable cooperation of these investigators, many of whom are renowned scientists who have worked in the field of ABE throughout their professional careers.

Lala I. P. Ray joins this book series as co-editor of this book volume. He is a frequent contributor to this book series and a staunch supporter of our profession.

The goal of this book volume, *Fertigation Technologies in Microirrigation: Crop Performance, Irrigation Requirements, Irrigation Efficiency,* is to guide the world science community on how judicious usage of irrigation water can put in practices by novel approaches.

We thank editorial staff and Ashish Kumar, Publisher and President at Apple Academic Press, Inc., for making every effort to publish the book when the diminishing resources are a major issue worldwide.

We express our deep admiration to our families for understanding and collaboration during the preparation of this book.

As an educator, there is a piece of advice to one and all in the world: "*Permit that our Almighty God, our Creator, allow us to inherit new technologies for a better life at our planet. We invite our community in agricultural engineering to contribute book chapters to the book series by getting married to our profession—*".

We are in total love with our profession by length, width, height, and depth. Are you?

—**Megh R. Goyal, PhD, P.E.**
—**Lala I. P. Ray, PhD**
Editors

PART I
Crop Performance under Micro Irrigation Systems

CHAPTER 1

PERFORMANCE OF DRIP-IRRIGATED TOMATO: WATER UPTAKE, ROOT DISTRIBUTION, AND QUALITY

RAJAN AGGARWAL, AJAY KUMAR VASHISHT, and NARINDER K. NARDA

ABSTRACT

This chapter focuses on the water uptake, root distribution, and quality aspects of tomato under drip and furrow methods of irrigation. The rooting density for furrow-irrigated treatment was higher in comparison with drip-irrigated plots for all depths. Specifically, the root concentration for both the furrow- and drip-irrigated plots was highest in the top 0–15 cm soil depth and was decreased with depth. In comparison to furrow-irrigated plots, the drip irrigation exhibited much higher values of specific water uptake during the early stages of growth. Also, the values of specific water uptake were independent of the depth of sampling. The values of various quality attributes of tomatoes (such as ascorbic acid, total soluble solids, and pH) were higher for drip-irrigated plots than those under furrow-irrigated treatments.

1.1 INTRODUCTION

Crop growth and its yield are influenced by the availability of water in the crop root zone. Soil moisture exercises its influence on the physiological processes of plant, which, in turn, governs the quality and yield of crop. The moisture is extracted from the root zone by plants through the root system, which extends or proliferates (as the case may be) to enable the plant to meet its consumptive use requirement.

Tomato (*Lycopersium esculentum* L.) is an important vegetable crop throughout the world. In Punjab, India, tomato is primarily grown as a winter–spring crop except for Hoshiarpur, Ropar, and Patiala districts, where it is grown as a rain-fed crop.[13] It has been estimated that the total production of tomato is 45 million metric tons from an area of 2.2 million ha.[16] Subject to the assured availability of irrigation water, tomato grows well in the cool and dry season.[10] Until today, the tomatoes are irrigated by furrow method of irrigation in areas having no shortage of water. In this system, soil water potential reaches up to saturated conditions (i.e., free water tension). Until the application of next irrigation, there is a gradual rise in the soil moisture tension.

Studies on the drip irrigation application on tomatoes[12,14] have revealed that maximum crop growth and yield were obtained by maintaining the moisture level of soil at a value very close to field capacity in the entire soil profile. As a result, the size of the plant and fruits increases, whereas the relative dry matter content, sugar, protein, and organic acid are decreased. The major function of the precise irrigation scheduling in tomato crop is to avoid water stress and to produce high yield with quality.[3,9,12]

Despite the progress made in evaluating irrigation scheduling of tomato crop in different regions, the farmers are still reluctant to deviate from their conventional irrigation methods. The widespread use of furrow irrigation method needs to characterize the moisture extraction level so that the crop does not suffer deleterious effects due to over/under irrigation.

The study of root development in time and space determines the quantity of water and nutrient reservoir available to a crop during its growth and the knowledge so gained can be used for better utilization of water and nutrients. Therefore, there is a need to evaluate the difference in the rooting behavior of tomato under the conventional furrow and drip irrigation. Hence, the first objective of this research was to compare the rooting distribution and water uptake patterns for furrow- and drip-irrigated tomatoes.

Further, it is apparent that the rooting pattern affects the quality of fruit. In the case of tomato, total soluble solids (TSS), ascorbic acid content, and acidity level govern the nutrient content and their utility for processing. To the best of authors' knowledge, no organized efforts have been made so far in determining these on a scientific basis and establish the qualitative changes by using various methods of irrigation. The second objective in this chapter was to compare various quality attributes of tomato under drip and furrow irrigation methods.

This chapter focuses on the water uptake, root distribution of tomato, and its quality aspects under drip and furrow methods of irrigation.

1.2 MATERIALS AND METHODS

An experiment was laid out at irrigation research farm of the Department of Soil and Water Engineering, College of Agricultural Engineering and Technology, Punjab Agricultural University, Ludhiana—India during 1991. Tomato seedlings of variety of "Punjab Kesari" were transplanted in the last week of December at six different plots, out of which three were drip irrigated (Fig. 1.1) and the rest three were furrow irrigated (Fig. 1.2). The values of field capacity and permanent wilting point were 0.17 and 0.05 cm^3/cm^3, respectively.[8] To protect the plants from frost damage during the early growing period, suitable plant protection measures were taken (Fig. 1.3).

FIGURE 1.1 Drip-irrigated plot after transplanting.

Irrigation was applied to both the treatments at 50% water extraction level. The daily potential evapotranspiration (PET) estimates were made by using modified Penman method. The 10-year daily meteorological data required for the study were procured from the meteorological laboratory of PAU. The average of daily PET values for the growing season of the crop was used for scheduling irrigation.

The daily crop evapotranspiration values are evaluated by multiplying daily crop coefficient values with PET estimate. The daily crop coefficient values are evaluated from crop coefficient curve, which was developed by using the meteorological data for the entire cropping season. The procedure outlined by Doorenbos and Pruitt[4] in FAO Irrigation and Drainage Paper 24 was employed to develop the crop coefficient curve.

FIGURE 1.2   Furrow-irrigated plots after transplanting.

FIGURE 1.3   Measures taken to protect plants from frost damage.

The soil moisture availability in the root zone at 15 cm depth interval was determined gravimetrically before and after the irrigation application. The depth of sampling was increased gradually as determined by soil profile exposures done on a fortnightly basis.

1.2.1 ROOTING DISTRIBUTION AND WATER UPTAKE

The objective was to determine the resultant differences in water uptake by the root system under both modes of irrigation so that proper water management strategies can be suggested to ensure better yield and quality of tomatoes.

To determine the rooting density, the root samples were taken from each plot with the help of a core sampler at 15 cm depth increment for the entire root zone of tomato crop at two locations (i.e., one adjacent to the plant and the other midway between the plants, respectively) (Fig. 1.4). The same procedure was followed for all plots. Before the core sampling, the penetration depth of the roots was monitored by exposing the soil profile (Fig. 1.5). The soil cores thus obtained were placed on a sieve and washed by repeatedly dipping and taking out sieve from a bucket full of water. After removing the foreign materials from each sample, the root length was measured by the relationship given by Tennant.[15]

FIGURE 1.4 Root sampling in drip-irrigated plots.

FIGURE 1.5   Determining rooting depth by exposing soil profile.

Root density measurements were taken throughout the growth period at an average interval of two weeks. In order to compare the water uptake patterns for the furrow- and drip-irrigated plants, volumetric soil moisture determinations were made for the drying cycles, which comprised rain less periods (following rain or irrigation during the growing season). Root density measurements were also carried out during these periods. Thus, the depletion of soil moisture during the entire drying cycle at any depth increment was estimated.

For the sake of simplicity, it was considered that the evaporation from soil is taking place only from the top 15 cm layer during the period of low-evaporative demand (usually prevalent during the crop season). Hence, to compare the water uptake from the top-most layer, the evaporation from soil was partitioned from transpiration.

Following empirical relationship[2] was used to estimate the magnitude of daily soil evaporation:

$$E_s = (1 - K_t) PET \times t^{-0.30} \tag{1.1}$$

where E_s is the daily soil evaporation (cm/day); K_t is the crop transpiration factor; and t is the time (day) after the soil was wetted last.

Earlier researchers have used the time exponent value equal to 0.50.[5,11] However, the rate and pattern of evaporation from bare medium textured soil exposed to low evaporation can be defined better through the selection of time exponent value equal to 0.30.[7] A maximum value of the factor K_t was 0.90 for the leaf area index (*LAI*) of >4.[6] However, K_t value does not decrease linearly with decrease in *LAI*. Therefore, for *LAI* <4, the following equation was considered so that the K_t values could decrease gradually:

$$K_t = 0.90 \times \left(\frac{LAI}{4.00} \right)^{0.50} \tag{1.2}$$

The amount of soil evaporation (E_s) estimated using eq 1.1 was subtracted from crop evapotranspiration for the drying cycle to get transpiration values. The estimated transpiration was then divided by the average root length density during the drying cycle to get the specific water uptake.

1.2.2   QUALITY ATTRIBUTES OF TOMATOES

- *TSS*: Ten mature tomatoes were randomly selected from each treatment to estimate TSS using band refractometer after correcting it for zero error. The observations were then averaged and expressed as degree brix at room temperature.
- *pH*: The pH values of tomatoes were determined with a digital pH meter. The observations were taken on five randomly selected tomatoes from each treatment and then averaged.
- *Ascorbic acid*: For estimating the ascorbic acid content in tomatoes, 2,6-dichloroindophenol method by AOAC[1] was used. The 2,6-dichloroindophenol ($C_{12}H_7Cl_2NO_2$) is used as a redox dye. When reduced, it becomes colorless. It imparts blue color in alkaline solution and the ascorbic acid reduced it to colorless solution. Ascorbic acid was estimated by taking five tomatoes from each treatment. A known weight of sample was added to 100 mL solution of 0.4% oxalic acid. It was filtered and 10 mL of filtrate was titrated against 2,6-dichloroindophenol dye. An appearance of pink color that persists for around 15 s indicates the end point. The results were expressed in mg of ascorbic acid per 100 g of tomatoes.

1.3   RESULTS AND DISCUSSION

The values of rooting density distribution for drip- and furrow-irrigated plots are presented in Table 1.1. It was revealed that the rooting density for furrow-irrigated plots (irrespective of the depth of sampling) was greater as compared to drip-irrigated plots. The values of rooting density however peaked at the same time (i.e., first week of April) and later were declined due to root senescence for both modes of irrigation. It was observed that the root concentration was decreasing with depth. However, it was highest in the top 0–15 cm soil layer.

TABLE 1.1   Comparison of Rooting Density for Furrow-Irrigated and Drip-Irrigated Tomatoes.

Days after transplanting	Position of root sampling	Depth increment (cm)	Root density (cm/cm^3)	
			Furrow	**Drip**
2	Near plant	0–15	0.34	0.16
18	Near plant	0–15	0.38	0.33
		15–30	0.05	0.03
	Between plants	0–15	0.15	0.10
33	Near plant	0–15	0.64	0.43
		15–30	0.12	0.12
	Between plants	0–15	0.33	0.12
		15–30	0.03	0.02
43	Near plant	0–15	0.78	0.51
		15–30	0.06	0.08
	Between plants	0–15	0.50	0.44
		15–30	0.04	0.04
60	Near plant	0–15	0.98	0.56
	Between plants	15–30	0.06	0.06
	Near plant	0–15	0.66	0.45
	Between plants	15–30	0.05	0.02
76	Near plant	0–15	1.20	0.68
		15–30	0.12	0.06
		30–45	0.03	0.02
	Between plants	0–15	1.31	0.53
		15–30	0.06	0.04
		30–45	0.02	0.02
89	Near plant	0–15	2.24	1.34
		15–30	0.32	0.11
		30–45	0.05	0.03
	Between plants	0–15	1.84	0.86
		15–30	0.18	0.08
		30–45	0.03	0.02

TABLE 1.1 *(Continued)*

Days after transplanting	Position of root sampling	Depth increment (cm)	Root density (cm/cm³)	
			Furrow	**Drip**
101	Near plant	0–15	3.11	1.85
		15–30	0.46	0.16
		30–45	0.09	0.06
	Between plants	0–15	3.00	1.23
		15–30	0.34	0.18
		30–45	0.06	0.05
113	Near plant	0–15	2.77	1.66
		15–30	0.34	0.14
		30–45	0.07	0.04
	Between plants	0–15	1.90	0.86
		15–30	0.24	0.13
		30–45	0.03	0.02
120	Near plant	0–15	1.83	0.93
		15–30	0.20	0.07
		30–45	0.06	0.04
	Between plants	0–15	0.93	0.27
		15–30	0.12	0.10
		30–45	0.03	0.02

This was probably due to the frequent wetting of top 0–15 cm layer in the case of drip-irrigated plots and relatively high frequency of irrigation in furrow-irrigated plots (50% water extraction level) so that most of the roots remained confined to the top layer. On the contrary, the rooting density was decreased as the sampling location moved laterally away from the plant. It was higher near the plant and less for a location mid-way between the plants for all depth increments in furrow- and drip-irrigated treatments.

Specific water uptake rates were estimated for furrow- and drip-irrigated plots (Table 1.2). It can be observed that for both the drip- and furrow-irrigated plots, the values of specific water uptake for the top 0–15 cm layer were lower as compared to 15–30 and 30–45 cm layers irrespective of the fact that whether sampling was carried out near the plant or in between the plants. It is logical because greater concentration of roots in the top 0–15 cm layer in comparison to bottom layers was observed and which, in turn, was responsible for higher extraction of water due to evaporation from the top 0–15 cm layer. Thus, a lower value of specific water uptake was expected in the top-most layer.

TABLE 1.2   Comparison of Specific Water Uptake for Furrow- and Drip-Irrigated Tomatoes.

Observation	Position of root sampling	Depth increment (cm)	Specific water uptake (cm³/cm)	
			Furrow	**Drip**
1	Near plant	0–15	0.13	0.35
2	Near plant	0–15	0.14	0.18
3	Near plant	0–15	0.07	0.16
	Near plant	15–30	0.33	0.97
	Between plants	0–15	0.09	0.20
4	Near plant	0–15	0.09	0.17
		15–30	0.18	0.48
	Between plants	0–15	0.11	0.23
5	Near plant	0–15	0.10	0.20
		15–30	0.25	0.88
	Between plants	0–15	0.14	0.29
		15–30	0.29	1.09
6	Near plant	0–15	0.08	0.09
	Between plants	15–30	0.44	0.49
	Near plant	0–15	0.10	0.11
	Between plants	15–30	0.56	0.70
7	Near plant	0–15	0.07	0.09
		15–30	0.44	0.85
	Between plants	0–15	0.09	0.11
		15–30	0.56	1.47
8	Near plant	0–15	0.07	0.08
		15–30	0.63	0.55
	Between plants	0–15	0.10	0.14
		15–30	0.91	0.69
9	Near plant	0–15	0.04	0.10
		15–30	0.37	0.69
		30–45	0.96	0.04
	Between plants	0–15	0.05	0.15
		15–30	0.53	0.91
		30–45	1.21	0.06
10	Near plant	0–15	0.02	0.05
		15–30	0.23	0.31
		30–45	0.77	0.51

TABLE 1.2 *(Continued)*

Observation	Position of root sampling	Depth increment (cm)	Specific water uptake (cm³/cm)	
			Furrow	Drip
	Between plants	0–15	0.03	0.06
		15–30	0.29	0.41
		30–45	0.96	0.64
11	Near plant	0–15	0.02	0.03
		15–30	0.18	0.17
		30–45	0.49	0.52
	Between plants	0–15	0.03	0.04
		15–30	0.23	0.22
		30–45	0.70	0.68
12	Near plant	0–15	0.02	0.03
		15–30	0.12	0.16
		30–45	0.11	0.18
	Between plants	0–15	0.02	0.05
		15–30	0.09	0.22
		30–45	0.14	0.22
13	Near plant	0–15	0.01	0.03
		15–30	0.04	0.18
		30–45	0.07	0.50
	Between plants	0–15	0.01	0.05
		15–30	0.06	0.24
		30–45	0.09	0.69
14	Near plant	0–15	0.01	0.02
		15–30	0.04	0.07
		30–45	0.09	0.10
	Between plants	0–15	0.01	0.03
		15–30	0.05	0.10
		30–45	0.13	0.12
15	Near plant	0–15	0.01	0.02
		15–30	0.06	0.04
		30–45	0.12	0.06
	Between plants	0–15	0.01	0.02
		15–30	0.08	0.05
		30–45	0.17	0.08

TABLE 1.2 *(Continued)*

Observation	Position of root sampling	Depth increment (cm)	Specific water uptake (cm³/cm)	
			Furrow	Drip
16	Near plant	0–15	–	0.01
		15–30	–	0.03
		30–45	–	0.08
	Between plants	0–15	–	0.02
		15–30	–	0.04
		30–45	–	0.10
17	Near plant	0–15	–	0.02
		15–30	–	0.05
		30–45	–	0.09
	Between plants	0–15	–	0.02
		15–30	–	0.07
		30–45	–	0.13
18	Near plant	0–15	–	0.05
		15–30	–	0.08
		30–45	–	0.17
	Between plants	0–15	–	0.07
		15–30	–	0.12
		30–45	–	0.14

Moreover, the value of specific water uptake was increased in deeper layers. This was probably due to the fact that gradual loss of water absorptive power of roots and their death due to aging in the top layer. The roots though less in number explored next deeper layers (i.e., 15–45 cm) for meeting the evaporative demand. For furrow-irrigated plots, specific water uptake rate was peaked for 0–15 cm layers in the first week of January, early-March for 15–30 cm layers, and middle-March for 30–45 cm layers. Thereafter, the specific water uptake rate was declined for all the layers. Also, the values of specific water uptake for a location in between the plants were higher as compared to sampling location near the plant due to less number of roots.

The drip-irrigated plot exhibited much higher values of specific water uptake during the early stages of growth as compared to the furrow-irrigated plots. The higher specific water uptake in drip-irrigated plots can also be attributed to the fact that the plants extracted water at relatively higher moisture level as compared to the furrow-irrigated plots.

It is an established fact that the soil moisture content decreases the capability of the soil to conduct water within the soil profile. Thus, in the case of furrow-irrigated plots, with diminishing soil moisture content within any drying cycle, the rooting system of the plant has to extract moisture by proliferating itself and so water extraction by roots also gets limited due to less availability of water. As a result, less water is transpired. Moreover, the plants through their physiological adaptation restrict the escape of water. However, in the case of drip-irrigated plots, the water is easily accessible to roots and is also transmitted quickly to the sites from where roots extract water. Hence, the loss of water is greater and consequently, the specific water uptake is also more.

The quality of tomatoes under the furrow- and drip-irrigated treatments was compared by considering ascorbic acid, TSS, and pH. It can be observed from Table 1.3 that the ascorbic acid for drip-irrigated tomatoes was higher than furrow-irrigated ones. Moreover, the similar trend was observed for TSS and pH also. Thus, as far as quality of tomatoes for processing is concerned, the drip-irrigated fruits had edge over furrow-irrigated fruits. With reference to quality aspect, ascorbic acid and TSS in tomatoes should have higher values and that was achieved in drip-irrigated tomatoes. Though high ascorbic acid is synonymous with high nutritive value, it can cause browning of Ketchup, which is a negative aspect in processing of tomatoes.

TABLE 1.3   Ascorbic Acid, TSS, and pH in Tomatoes under all the Treatments.

Analysis no.	Furrow			Drip		
	Ascorbic acid	TSS	pH	Ascorbic acid	TSS	pH
	mg/100 g	°Brix	–	mg/100 g	°Brix	–
1	24.96	3.81	4.13	28.08	4.70	4.15
2	23.40	3.81	4.26	29.84	4.20	4.44
3	29.25	4.27	4.24	26.91	4.33	4.30
4	31.98	4.67	4.33	31.00	4.90	4.25
5	28.60	4.13	4.18	29.83	4.30	4.23
6	25.74	4.00	4.22	30.81	4.67	4.29
7	28.66	4.61	4.33	30.42	4.90	4.25
8	28.27	4.20	4.27	28.73	4.67	4.31
9	27.10	4.60	4.32	28.08	5.13	4.29
10	26.91	4.42	4.26	28.08	4.70	4.30
11	27.49	4.21	4.26	29.18	4.65	4.28

Higher value of TSS means that lesser quantity of tomatoes will be needed for processing. Moreover, higher TSS value is also known to improve the taste of tomatoes. Hence, higher values of ascorbic acid and TSS achieved in drip-irrigated tomatoes as compared to furrow-irrigated ones favor the adoption of drip irrigation technology for raising tomatoes for processing purposes. The only limitation seems to be higher pH, which was attained in drip-irrigated tomatoes. It should be less than 4.25, whereas it was found to be marginally higher.

Following conclusions are drawn from this research study:

- From the field experimentation, it has been concluded that the rooting density for furrow-irrigated treatment was greater in comparison to drip-irrigated plots for all depths. Specifically, the root concentration for both the furrow- and drip-irrigated plots remained highest in the top 0–15 cm layer and decreased with depth.
- In comparison to furrow-irrigated plots, the drip-irrigated ones exhibited much higher values of specific water uptake during the early stages of growth. It is also concluded that the values of specific water uptake were independent of depth of sampling.
- It was also concluded that the values of various quality attributes of tomatoes (such as ascorbic acid, TSS, and pH) were higher for drip-irrigated plots than all the furrow-irrigated treatments.

1.4 SUMMARY

In this study, rooting density for furrow-irrigated treatment was greater as compared to drip-irrigated plots for all depths. The values of rooting densities for all irrigation treatments, however, peaked in the first week of April. The rooting concentration for both the drip and furrow-irrigated plots remained highest in the top 0–15 cm soil layer. In general, the rooting concentration was decreased with depth. The drip-irrigated plots exhibited much higher values of specific water uptake during the early stages of growth as compared to furrow-irrigated plots. The values of specific water uptake for 0–15 cm layers was lower as compared to 15–30 cm layers, irrespective of the fact that whether sampling was carried out near the plant or in between the plants. Also, the values of specific water uptake were higher for the sampling done near the plant as compared to sampling done in between plants (irrespective of depth of sampling).

The ascorbic acid, TSS, and pH values for tomatoes under drip irrigation were higher than all furrow-irrigated treatments. The results strongly favor the adoption of drip irrigation technology for raising tomatoes for processing purposes.

KEYWORDS

- **crop transpiration factor**
- **drip irrigation**
- **furrow irrigation**
- **leaf area index**
- **potential evapotranspiration**

REFERENCES

1. Anonymous. *Official Method of Analysis,* 11th ed.; Association of Official Agricultural Chemists (AOAC): Washington, DC, 1970; p 777.
2. Arora, V. K.; Prihar, S. S.; Gajri, P. R. Synthesis of a Simplified Water Use Simulation Model for Predicting Wheat Yields. *Water Res. Res.* **1987,** *23* (5), 903–910.
3. Cannell, G. H.; Asbell, C. W. Irrigation of Field Tomatoes and Measurement Soil Water Changes by Neutron Moderation Methods. *J. Am. Soc. Hortic. Sci.* **1974,** *96,* 305–308.
4. Doorenbos, J.; Pruitt, W. O. (Eds.). *Guidelines for Predicting Crop Water Requirements.* FAO Irrigation and Drainage Paper 24, (Rev.); FAO Food & Agricultural Organization: Rome, 1977; p 156.
5. Hanks, R. J. Model for Predicting Plant Yield as Influenced by Water Use. *Agron. J.* **1974,** 66, 660–665.
6. Hanks, R. J.; Puckridge, D. W. Prediction of the Influence of Water, Sowing Data and Planting Density on Dry Matter Production of Wheat. *Aus. J. Agric. Res.* **1980,** *31,* 3–11.
7. Jalota, S. K.; Parihar, S. S. Effects of Atmospheric Evaporativity, Soil Type and Redistribution Time on Evaporation from Bare Soil. *Aust. J. Soil Res.* **1986,** *24,* 357–366.
8. Malik, R. K. Development of Root Growth and Water Uptake Model for Wheat. Ph.D. Thesis, Punjab Agricultural University, Ludhiana, 1985; p 289.
9. Phene, C. J.; McCormick, R. L. In *Evapotranspiration and Crop Coefficient of Trickle Irrigated Tomatoes,* Proceeding of Third International Drip/Trickle Irrigation Congress, Fresno, CA; American Society of Agricultural Engineers: St. Joseph, MI, 1985; pp 823–831.
10. Ramalan, A. A.; Nwokeocha, C. U. Effects of Furrow Irrigation Methods, Mulching and Soil Water Suction on the Growth, Yield and Water Use Efficiency of Tomato in the Nigerian Savanna. *Agric. Water Manage.* **2000,** *45* (3), 317–330.

11. Ritchie, J. T. Model for Predicting Evaporation from a Row Crop with Incomplete Crop Cover. *Water Resour. Res.* **1972,** *8*, 1204–1213.

12. Rudich, J.; Kalmar, D.; Geizenberg, C.; Harel, S. Low Water Tensions in Defined Growth Stages of Processing Tomato Plants and Their Effects on Yield and Quality. *J. Hortic. Sci.* **1977,** *52* (3), 391–399.

13. Sharma, J. R. Studies on the Adaptability of Some Tomato Varieties in the State of Punjab. Ph.D. Thesis, Department of Vegetable Crops, Landscaping and Floriculture, Punjab Agricultural University, Ludhiana, India, 1981; p 250.

14. Shrivastava, P. K.; Parikh, M. M.; Sawani, N. G.; Raman, S. Effect of Drip Irrigation and Mulching on Tomato Yield. *Agric. Water Manage.* **1994,** *25* (2), 179–184.

15. Tennant, D. Test of a Modified Line Intersect Method of Estimating Root Length. *J. Ecol.* **1975,** *63*, 995–1001.

16. Villareal, R. (Ed.). *Tomato in the Tropics*; Westview Press: Boulder, CO, 1980; p 174.

CHAPTER 2

PERFORMANCE OF CUCUMBER UNDER MICRO- AND MINI-SPRINKLER IRRIGATION WITH LAND SLOPES

R. CHELLENG, ANAMIKA YADAV, LAXMI NARAYAN SETHI, SUDIPTO SARKAR, and AVINASH KUMAR

ABSTRACT

Water saving irrigation method in the North-East region of India is a better option for high yield and crop productivity. Initially, two replicate field plots in the hilly terrain of the Department of Agricultural Engineering, Assam University, Silchar were used to evaluate yield, water-use efficiency (WUE), and economic analysis of cucumber cultivation under J-master mini sprinkler (T_1) and Aqua-master micro-sprinkler (T_2). Slope of 20% (plot I) and 40% (plot II) was considered for the experimental plots to determine any variants affecting the yield and WUE. The growth and yield parameters of cucumber under Aqua-master micro-sprinkler irrigation resulted in higher vegetative growth (plant vine length of 51.09%) and yield (number of fruits of 16.67% and total yield of 0.26%) than the plants under J-master mini sprinkler irrigation. Benefit–cost ratio for T_2 was higher than T_1 in both plots. It is concluded that the farmers can get higher net returns for cucumber with Aqua-master sprinkler irrigation. The payback period for the production of cucumber with J-master and Aqua-master sprinklers can be achieved in less time compared to J-mini sprinkler.

2.1 INTRODUCTION

Irrigation practices in agricultural field plays a vital role for food production and enhancement of the effective water use per unit the actual water withdrawal. Water stress can be detrimental to the overall turnout of the growing plant.[19,37]

The drip and sprinkler irrigation systems are key interventions in water saving and improving crop productivity.[6,7,32] Many of the research studies in India have indicated that 40–80% of water saving and 100% yield increase with micro irrigation.[40] The micro irrigation system has become popular in intensive water scarcity regions for horticultural and commercial crops.

The world's wettest region, North-East of India, has an average annual rainfall of 2000 mm, and about 72% of area is considered as hilly; therefore, incoming rainfall run away to the downward slope, thus affecting crop growth due to low water availability. In the hilly areas, shifting cultivation is a primitive practice for the excess water unneeded except for the tea plantations and other agricultural crops on terraces and hilltops, and irrigational facilities are mandatory. Therefore, sprinkler irrigation is an alternative for water saving and fertigation.[21] The research study in winter season indicated higher crop production rate under sprinkler irrigation than the surface irrigation.[43] It was concluded that micro irrigation is a paramount management practice to reduce the water scarcity issues.[28] Horticultural production and product exportation are increasing firmly because of market demand of developing nations, though the yield has reduced.[45]

Almost 0.575 million ha of area is under horticulture crops in the state of Assam, which is 14.04% area out of total cultivable area of 4.099 million ha in 2011–2012. Appropriate management of horticultural crops with novel scientific methods can enhance the rural economy to provide the nutritional security.[17]

The Cachar district of Assam has an elevation of 36.5 m above MSL (longitude of 92°24′E and 93°15′E and latitude of 24°22′N and 25°08′N). The entire Cachar district is surrounded from three sides by hill ranges of Manipur, Mizoram, and Meghalaya. It is categorized as undulating topography, uneven hilly terrain, mounts, large piles, and depressed waterlogged areas called *beels*. The general elevation of the land surface ranges from 450 to 1100 m. The humid subtropical climate dominates in this region. Through the "tropical monsoon rainforest climate," this region can be considered as moderate, heavy rainfall with high humidity, while the distribution is not uniform.[11,38] The exclusive agroclimatic condition by well-drained fertile soil is favorable for growing various varieties of horticulture crops.[17] Assam hills have abundant land resources compared to fewer water supplies for irrigation, which make an issue to get sufficient soil capacity. Therefore, implementation of sprinkler irrigation is required.

Cucumber (*Cucumis sativus* L.) belongs to *Cucurbitaceous* family.[1,30] Economically cucumber is ranked fourth after tomatoes, cabbage, and onion in Asia.[13] It is a trailing crop that is usually grown for the fruits to be eaten raw, cooked, or fried and for numerous therapeutic properties.[26] The other uses include culinary and alternative medicine.[18,29] Cucumber is very good for health and contains water, soluble fibers, vitamins (A, C, K, and B_6), and potassium.[44] High temperatures and sufficient soil moisture are favorable for high production, while the unfavorable climatic conditions can reduce the growth of plant,[10] fruits[22,24,25] and cause mineral disorder.[4] Thus, the planting must be done in the spring–summer season to achieve higher yield.

Globally, India occupies the largest area under irrigation.[35] The irrigation process is application of water uniformly over the land to protect the soil structure, which stores and hinders the natural flow. This method helps one to protect plants against frost and unwanted weed growth.[39,41] Mostly 70% of fresh water is used for irrigation for food production.[9,15] Approximately 18% of the total cropped land produced 40–45% of the food worldwide in 2003.[12] The major goal of irrigation is to supply sufficient quantity of water.[33] Therefore, an adequate amount of water at a proper time can increase the yield, while the scarcity might be the reason for low quality of crop.[37]

This chapter focuses on the potential and adaptability of *C. sativa* L. cultivation; performance evaluation (plant yield, WUE, and economical parameters) in hilly terrains under different micro-sprinkler irrigation systems in a hillock.

2.2 MATERIAL AND METHODS

2.2.1 EXPERIMENTAL SITE

The experimental site was a slopping land area in the hilly terrains of Assam University, Silchar (24.68°N, 92.75°E) in the North-East region of India. The hilly terrains comprise undulating topography called *beels*. The experimental site is shown in Figures 2.1 and 2.2.

2.2.2 FAVORABLE SITES FOR CROP PLANTATION

2.2.2.1 TOPOGRAPHY OF LAND

Natural slope is a measure the proficiency of proper land for agriculture uses.[5,16,31,42] In pursuance of British land capability, slope can directly affect cultivation practices of crop. For example, >15–20° slope is unfit for tillable crops and >20° slope is usually tricky for ploughing and fertilization operations.

Generally, steep slopes are not efficient for agriculture production, which is associated with high risk for soil degradation and landslides.[3,8,14]

Terrain refers to the quantitative measurement of vertical elevation change in cropland area, which influences the potential use and management of agricultural machines.[36] The influence of terrain on soil erosion is more important rather than other risk factors, including climate, soil type, crop type, and preventive methods of erosion.[2,27] For integrated crop production, mechanization helps one to intensify the production in a sustainable manner, while accessibility of agricultural field activities takes a lot of time.[20]

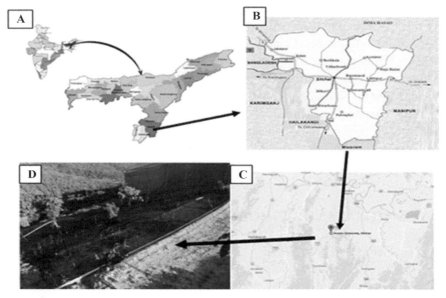

FIGURE 2.1   Location of the experimental site: (A) Map of India; (B) Assam State; (C) The hilly terrain of Assam University, Silchar; and (D) Selected field site.

2.2.2.2   SOIL TYPE

The soil found in Barak Valley Zone is of sedimentary rock, with texture from sandy to clayey. Most of the area is covered with old riverine alluvium soil, red soil, and peat soil. Extreme care is required to maintain the soil fertility.[34] Cucumber crop is well grown in light-textured soils but will produce early in sandy soil with well drainage, high organic matter, and pH range of 6–6.8.

FIGURE 2.2   Aerial view of a sloping land at the experimental site: (A) and (B).

2.2.2.3   CLIMATIC CONDITIONS

The Barak Valley zone is subtropical that is characterized by moderately hot and moist climate. The average annual precipitation of Silchar is about 2196 mm and sometimes accompanied with hails. Season to season, the rainfall pattern varies from low and erratic rainfall during March–April and high rainfall during May–September. The region experiences summers during the months of April–August with a maximum temperature of 32°C and winter during the months of November–January with a minimum temperature of around 11°C. The monsoon season in Silchar is from July to September. The cucumber is an easy-care vegetable crop that loves warm weather condition (27–30°C) and plenty of sunlight. The most effective environmental weather parameter is temperature, which can influence the growth of vegetation, flower initiation, fruit growth, and fruit quality.

2.2.3   MINI SPRINKLER IRRIGATION FOR CROP PLANTATION

The sprinkler system is considered for water rescue by using it at proper time and quantity as compared to surface irrigation.[21] This system distributes water in the form of spray through the network of pipes and sprinklers, somewhat resembling as a natural rain. With this method, crop can be irrigated satisfactorily on soils having very high intake rate, steep slopes, and irregular topography.

The J-master mini and Aqua-master micro-sprinklers were selected for this study based on the equipment cost, labor requirement, and suitability for the particular farming operations. In mini sprinkler irrigation under pressure, water is sprayed out in the form of rain; hence, it simulates rainfall by uniform application of water. This mini sprinkler should be adopted under low adjustable discharge rates for many close-growing crops; thus, it is a water saving method compared with conventional method. With this method, the crop quality and quantity can be improved against fluctuating weather conditions. The overall performance of the system depends on the uniformity distribution of water with proper and efficient way.[23]

Micro-sprinkler irrigation is the straight application of water on the soil surface to dissipate initially under low pressure and then permeate into the wetted profile, through uniformly distribution of water in orchards. For example, fixed-head sprinklers are useful for small-area irrigation like lawns and gardens and another perforated pipes system can also be operated at low pressure and releases more water per unit time rather than rotating head sprinklers. It is an efficient and alternative method of irrigation for high-value cash crops. With proper management, this method can increase yield with decreasing the demand for water use and fertilizer requirements. It is ideal for sloppy or irregularly shaped orchard blocks that cannot be flooded or furrow irrigated.

2.2.4   EXPERIMENTAL SETUP

The experiment was conducted in two plots of different slopes situated in the hilly terrain of the Department of Agricultural Engineering. The total area covered by the plots was 64 m² (32 m² for each plot). Both experimental plots were further divided into two sections for two treatments having an area of 16 m² each (Fig. 2.3). The discharge rate, the size of laterals and mains, operating pressure, and water application rate were designed using eqs 2.1–2.8 and the values are given in Table 2.1.

The row-to-row and plant-to-plant distances were 8 m × 4 m per plot. Therefore, four J-master mini sprinkler and one Aqua-master micro-sprinkler were used in each plot. The discharge, size of laterals and mains, and operating pressure for mini and micro-sprinkler was 7.6 and 10.10 L/h; 16 and 50 mm; 98 and 196 kPa, respectively. The major head loss was neglected for both sprinklers as the lateral length was only 2.70 and 2.89 m, respectively. The application rate and capacity of J-master mini sprinkler and Aqua-master micro-sprinkler were 0.232 cm/h and 0.522 m³/h, respectively.

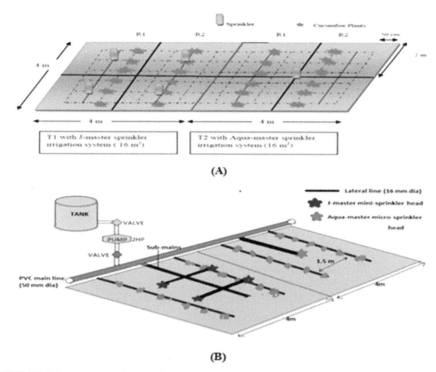

FIGURE 2.3   The experimental layouts: (A) and (B).

TABLE 2.1   Design Parameters for Micro irrigation Systems.

Design parameter	Unit	J-master mini sprinkler		Aqua-master micro-sprinkler	
		Value	No.	Value	No.
Sprinkler discharge	L/h	16	4	30	1
Size of laterals	mm	16	4	16	1
Main line	mm	90	1	90	1
Submain line	mm	50	1	196	1
Operating pressure	kPa	98	4	1.04	1
Discharge rate	m³/h	1.986	4	0.464	1
Application rate	cm/h	1.985	–	30	–

2.2.5   PERFORMANCE EQUATIONS FOR IRRIGATION METHODS

In order to assess the performance of the sprinkler systems, discharge rate, uniformity coefficient of each emitter, and wetted diameter for each sprinkler were evaluated with eqs 2.1–2.8. The uniformity coefficient was calculated using the following formula:

$$CU = \left\{ 1 - \left(\frac{\sum x}{mn} \right) \right\} \times 100 \qquad (2.1)$$

where CU is the coefficient of uniformity; n is the number of observations; m is the average catch; $\bar{x} = |z - m|$ is the deviation from mean.

The operating pressure of sprinklers can be defined by the standard pressure formula:

$$P = \rho g h \qquad (2.2)$$

where P is the pressure (Pa); ρ is the water density (1000 kg/m³); g is the gravitational acceleration = 9.81 m/s²; and h is the height = 8 m in this study.

Sprinkler irrigation releases water frequently depending on daily need basis. Therefore, water required per day water was estimated as follows:

$$WR \ (L/day) = ET \times KC \times Cp \times Area \qquad (2.3)$$

where ET is the evapotranspiration (mm/day); Kc is the crop factor; and Cp is the canopy factor area (m²).

The capacity of each sprinkler was calculated from the following formula:

$$Q = S_1 Smr \qquad (2.4)$$

where Q is the discharge of each sprinkler (m³/h); S_1 is the sprinkler spacing along the lateral (m); Sm is the sprinkler spacing between lines or along the main (m); and r is the rate of application (m/h).

The application rate can be calculated by the following equation:

$$\text{Application rate (cm/2h)} = \left(\frac{\text{discharge of sprinkler} (L/s) \times 360}{\text{sprinkler spacing} (m) \times \text{lateral move} (m)} \right) \qquad (2.5)$$

Irrigation time can be calculated as follows:

$$\text{Irrigation duration (h/day)} = \frac{\text{Water requirement} (L/day)}{\text{Application rate } (L/h)} \qquad (2.6)$$

WUE and irrigation water-use efficiency (IWUE) can be determined according to the following formulas:[19]

$$WUE = \left(\frac{E_Y}{E_t} \right) \times 100 \qquad (2.7)$$

where WUE is the water-use efficiency (t/ha mm); Ey is the economical yield (t/ha); and Et is the plant water consumption (mm).

$$\text{IWUE} = \left(\frac{E_Y}{I_r}\right) \times 100 \tag{2.8}$$

where IWUE is the irrigation water-use efficiency (t/ha mm); Ir is the amount of applied irrigation water (mm).

2.3   FIELD EXPERIMENTATION

2.3.1   STATUS OF SOIL PROPERTIES AND NUTRIENT APPLICATION

In order to carry out the field experiment, the soil physicochemical characteristics (such as soil texture, bulk density, moisture content, soil pH, and electrical conductivity [EC]) were determined using standard methods (Table 2.2).

TABLE 2.2   Average Nutrient Content of Soil in the Plots at Initial Status.

Soil depth	Average initial nutrient content in the soil			
	Total nitrogen	Total phosphorus	Total potassium	Total organic carbon
cm		mg		%
Field plot I				
0–15	134	1.91	79	0.49
15–30	126	1.24	55	0.45
30–45	104	2	55.5	0.51
45–60	125	1.725	56.25	0.555
Average	14.65	0.34	17.80	0.15
SD	134	1.91	79	0.49
Field plot II				
0–15	116	0.75	36.7	0.79
15–30	114	1.61	59	0.59
30–45	127	1.04	65	0.55
45–60	154	1.2	54.5	0.51
Average	127.75	1.15	53.8	0.61
SD	18.41	0.36	12.18	0.12

2.3.2   GROWTH AND AGRICULTURAL PRODUCTION AND COST—ECONOMICS

The classical approach in plant growth analysis was used to find out the relative growth rate with plant weight at different harvest times. The growth parameters were number of branches, leaves, flowers, fruits at the time of flowering and harvesting, fruit length and width, fruit weight, and yield per plant. The cost of production components was cost for land preparation, nursery and seedlings preparation, manures and fertilizers, plant protection measures, labor cost, land revenue, etc. The production cost components of cucumber cultivation were gross returns, net returns, net profit, benefit–cost ratio, and payback period.

2.4   RESULTS AND DISCUSSION

2.4.1   PHYSICAL CHARACTERISTICS OF SOIL

Soil samples were collected from 16 observation locations (eight samples from each plot) at four soil depths (0–15, 15–30, 30–45, and 45–60 cm) to analyze the soil physicochemical and nutrient properties at the experimental site.

2.4.1.1   TEXTURAL CLASSIFICATION

The soil texture based on the USDA textural classification is shown in Figure 2.4. The sand, silt, and clay percentages at four soil depths in both the plots are shown in Figure 2.5. Figures 2.4 and 2.5 indicate that the soil at a depth of 0–15 cm was loamy sand; sandy loam for soil depth 15–30 cm, while for depths 30–45 and 45–60 cm it was sandy clay loam in plots I and II, respectively.

The percentage of sand was highest among all aggregates in both treatments. Based on the USDA textural triangle, it was observed that the soil in plots I and II was sandy clay loam with sand, silt, and clay percentage of 68.2, 8, and 27.3% in plot I and 64.31, 8.75, and 26.9% in plot II, respectively.

2.4.1.2   SOIL MOISTURE CHARACTERISTICS

For the estimation of soil water characteristics, the soil samples were collected from four soil depths at different tensions (0.3, 1, 5, and 15 bars) using pressure plate apparatus and hot air oven. The soil moisture at four depths is shown in Table 2.3 in both treatments.

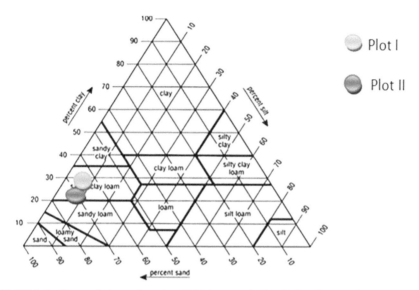

FIGURE 2.4   Textural classes based on USDA textural triangle for plots I and II.

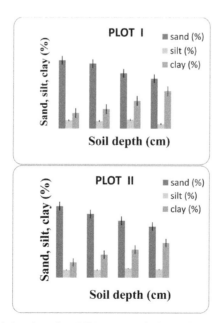

FIGURE 2.5   Textural class based on USDA textural triangle in plots I and II.

TABLE 2.3   Soil Moisture Characteristics at Varying Soil Depths.

Soil depth (cm)	Soil moisture content (%) at different tensions (bar)							
	0.3	1	5	15	0.3	1	5	15
	Plot I				Plot II			
0–15	21.76	18.42	17.52	16.79	20.06	17.05	15.56	15.26
15–30	19.49	17.85	16.47	15.21	18.37	16.32	15.26	14.27
30–45	18.52	17.02	15.72	14.51	18.09	15.56	14.25	13.28
45–60	18.04	16.49	15.06	13.95	17.91	15.03	12.82	12.43

Table 2.3 and Figures 2.6 and 2.7 indicate that the amount of soil moisture at field capacity (FC) and wilting condition was in the decreasing order with soil depth in both treatments due to variable topography and land use, which affects the soil moisture distribution and management practices. The soil water content at FC in plots I and II was maximum at depth 0–15 cm (21.76 and 20.06%), respectively, and was minimum at depth of 45–60 cm (13.95 and 12.82%), respectively. The wilting point for plots I and II was maximum at soil depth of 0–15 cm (16.79 and 15.26%), respectively, and was minimum at soil depth of 45–60 cm (13.95 and 12.43%), respectively.

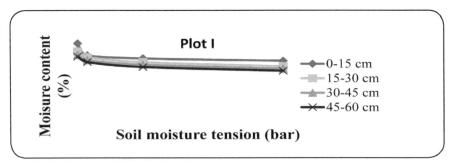

FIGURE 2.6   Soil moisture status (on dry basis) at four soil depths: plot I.

2.4.2 PHYSIOCHEMICAL PROPERTIES OF SOIL

2.4.2.1 pH VALUE AND ELECTRICAL CONDUCTIVITY

The pH and EC of soil were determined by an electrometric method using pH meter and a conductivity meter, respectively. The observed pH and EC values in plots I and II are given in Figure 2.8. The soil pH value in both treatments ranged from 4.26 to 5.82, thus reflecting acidity of soil. Figure 2.8

shows the variation of average pH at four soil profiles in both the plots. The pH was higher at a depth of 45–60 cm for both plots compared to other soil layers. The EC value ranged from 0.02 to 0.49 mS. Figure 2.8 shows the variation of the average EC at soil profiles in both the plots. The maximum EC value was at the depth of 30–45 cm in both plots followed by 15–30 cm soil profile.

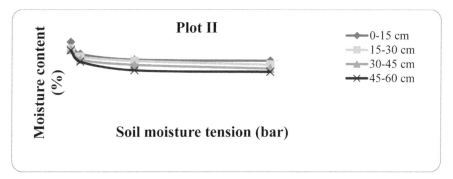

FIGURE 2.7   Soil moisture status (on dry basis) at four soil depths: plot II.

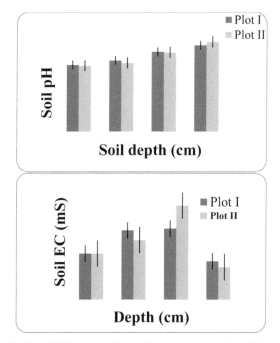

FIGURE 2.8   Soil pH and EC status at four soil depths in plots I and II.

2.4.3 EFFECTS ON GROWTH PARAMETERS

The vegetative growth parameters (vine length and leafage bulk per plant; plant height at 10-day interval) and yield parameters (number of fruits, fruit density, and total yield) were monitored for mini- and micro-sprinkler irrigation systems.

2.4.3.1 EFFECT ON VINE LENGTH PER ROW OF CUCUMBER PLANT

The vine length per plant of cucumber is shown in Table 2.4 and Figures 2.9 and 2.10. The effects of irrigation were significant ($p < 0.05$) on vine length at different growth stages. The J-master mini sprinkler irrigation resulted in plants with the longest vine length at all growth stages compared to Aqua-master micro-sprinkler irrigation.

TABLE 2.4 Variation in Average Vine Length of Cucumber in Plots I and II.

Days after plantingays	Vine length per plant, mm							
	J-Master mini sprinkler				Aqua-master micro-sprinkler			
	R_1	R_2	Average	SD	R_3	R_4	Average	SD
			Plot I					
30	322	280	301	30	365	370	368	4
40	720	725	723	4	820	792	806	20
50	945	1028	987	59	1177	1208	1193	22
60	1430	1410	1420	14	1547	1550	1549	2
			Plot II					
30	358	365	361	5.2	287	322	304	24
40	838	820	829	13.0	623	578	601	31
50	1188	1177	1183	8.2	892	912	902	14
60	1387	1547	1467	113.1	1177	1430	1303	17

R, replication.

2.4.3.2 EFFECT ON NUMBER OF FLOWERS PER PLANT OF CUCUMBER

The data for the effects on number of flowers and number of fruits per plant are presented in Tables 2.5 and 2.6, and Figures 2.11 and 2.12, respectively, for both treatments (T_1 and T_2).

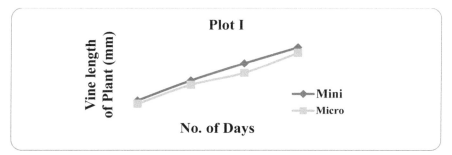

FIGURE 2.9   Variation in average vine length (mm) of cucumber in plot I.

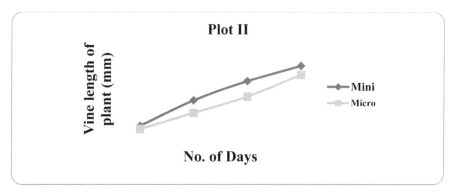

FIGURE 2.10   Variation in average vine length (mm) of cucumber in plot II.

TABLE 2.5   Variation in Number of Flowers per Plant in Each Treatment.

Days after planting, days	No. of flowers per plant							
	J-Master mini sprinkler				Aqua-master micro-sprinkler			
	R_1	R_2	Average	SD	R_3	R_4	Average	SD
	Plot I							
40	7	7	7	0	6	6	6	0.28
50	17	16	16	0.57	14	15	15	0.57
60	23	27	25	2.40	23	24	23	0.57
	Plot II							
40	5	5	5	0.24	6	5	5	0.35
50	10	9	10	0.59	8	9	8	1.00
60	18	18	18	0.00	18	16	17	1.53

TABLE 2.6   Variation in Number of Fruits per Plant in Each Treatment.

Harvest	No. of fruits per plant			
	J-master sprinkler, T_1		Aqua-master sprinkler, T_2	
	R_1	R_2	R_3	R_4
	Plot I			
1st harvest	22	19	15	17
2nd harvest	17	14	14	12
Average	19.5	16.5	14.5	14.5
Standard deviation	3.54	3.54	0.71	3.54
	Plot II			
1st harvest	18	17	16	15
2nd harvest	14	13	12	10
Average	16	15	14	12.5
Standard deviation	2.83	2.83	2.83	3.54

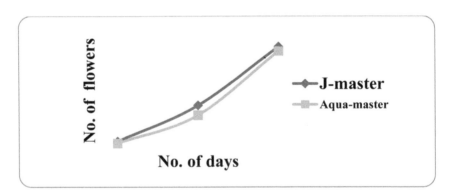

FIGURE 2.11   Variation in number of flowers of cucumber plant for plot I.

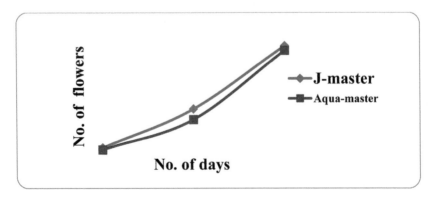

FIGURE 2.12   Variation in number of flowers per cucumber in plot II.

The number of flowers at 10-day interval after 60 days of planting was recorded in plot I and the treatment T_1 (R_2) recorded 27 flowers per plant. The lowest number of branches per plant was 22 in treatment T_2 (R_3) in plot II. The treatment T_1 (R_2) recorded 18 flowers per plant and the least number of branches per plant was 16 in treatment T_2 (R_3).

2.4.3.3 EFFECT ON CUCUMBER YIELD PER PLANT

The variation of fruit numbers and fruit weight for both treatments are given in Table 2.7. Figure 2.13 shows the cucumber crop at maturity and harvest stages. The total yield in plot I was highest in T_1 (18.8 kg) than in the T_2 (17.2 kg). Similarly in plot II, the total yield was highest in T_1 (13.6 kg) than in the T_2 (12.7 kg) (Table 2.7).

FIGURE 2.13 Cucumber yield: (a) Maturity stage; (b) Harvested fruits.

Total yield per plant was 36.2 kg in plot I and 26.3 kg in plot II, respectively (Table 2.7). No significant differences were observed in both treatments at the same temperature and humidity. The slope aspects caused greater differences in yield on plot basis. Figure 2.14 shows variation of fruit weight throughout the growing season and variation of total yield per replication in both treatments in both the plots. The variation in fruit weight throughout the growing season in both treatments is illustrated in Figure 2.15.

TABLE 2.7 Variation in Fruit Weight throughout the Growing Season in Plots I and II.

Harvesting stage	Fruit weight						Total yield (kg)
	J-master sprinkler, T_1			Aqua-master sprinkler, T_2			
	R_1, g	R_2, g	Total yield, kg $R_1 + R_2$	R_3, g	R_4, g	Total yield, kg $R_3 + R_4$	
Plot I							
1st harvest, **A**	5381.5	4823.4	10.2	4331.4	4789.4	9.1	36.0 = **C** + **D**
2nd harvest, **B**	4487.6	4203.5	8.6	4156.4	4005.6	8.1	
Total (kg), **A** + **B**	9.9	9.0	18.8, **C**	8.5	9.0	17.2, **D**	
Avg. (kg)	4.9	4.5	–	4.2	4.5	–	–
Plot II							
1st harvest, **A**	3881.6	3651.1	7.5	3534.5	3489.5	7.0	26.3 = **C** + **D**
2nd harvest, **B**	2987.7	3163.2	6.1	2906.5	2865.7	5.7	
Total (kg), **A**+**B**	6.9	6.8	13.6, **C**	6.4	6.4	12.7, **D**	
Avg. (kg)	3.43	3.41	–	3.22	3.18	–	–

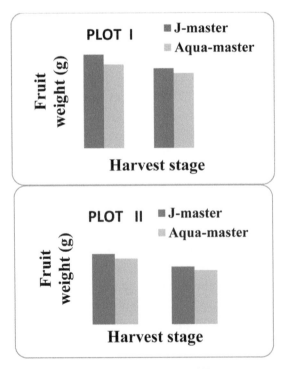

FIGURE 2.14 Variation in fruit weight (g) in plots I and II.

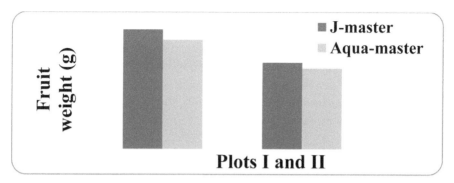

FIGURE 2.15   Comparison of fruit weight (g) between plots I and II.

2.4.4   ESTIMATION OF WATER-USE EFFICIENCY

For the estimation of the WUE, equal quantity of water was applied from both irrigation systems at the same time. There was no significant difference in WUE between the treatments of the same plot but variation was observed for different slopes. As shown in Table 2.8 and Figure 2.16, the highest value was 5.30 t/ha-mm in T_1 for J-master sprinkler irrigation system for plot I, while the lowest value was 3.59 t/ha-mm in T_2. Similarly, the highest IWUE value was 4.78 t/ha-mm in T_1 for J-master sprinkler irrigation in plot I, while the lowest value was 3.24 t/ha-mm in T_2. It revealed that WUE and IWUE for both treatments do not differ significantly, but the slope may affect the efficiency.

TABLE 2.8   The Variation in Water-Use Efficiency of Cucumber in Plots I and II.

Parameter	Unit	Plot I		Plot II	
		J-master, T_1 sprinkler)	Aqua-master, T_2 sprinkle	J-master, T_1 sprinkler)	Aqua-master, T_2 sprinkler)
Economic yield	t/ha	11.8	10.8	8.6	8
Water consumption	mm	223	223	223	223
WUE	t/ha -mm	5.30	4.85	3.84	3.59
Irrigation applied	mm	247	247	247	247
IWUE	t/ha mm	4.78	4.38	3.47	3.24

2.4.5   COST ECONOMICS OF CUCUMBER PRODUCTION

The cost of production of cucumber in the two plots under two irrigation systems was estimated considering different components of cost of production (Table 2.9). The unit cost for each component was collected from the market and used for the analysis of production cost. The cost of production does not include the cost involved in setting up the irrigation systems. Crop production data were also collected for the cucumber crop grown from March 2017 through May 2017. Among all the components for the total cost of production of cucumber in both the plots for two treatments, the hired labor cost was maximum, followed by the cost of manure and fertilizers and land revenue, etc.

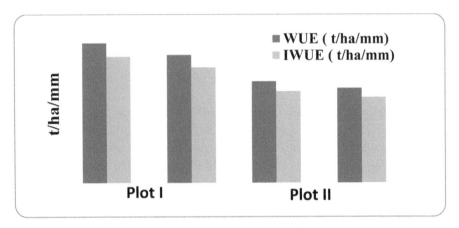

FIGURE 2.16   Water-use efficiency of cucumber for plots I and II.

TABLE 2.9   Cost of Production of Cucumber.

Component	Cost of production of cucumber per plot (Rs.)			
	Plot I		Plot II	
	T₁ (mini-sprinkler)	T₂ (micro-sprinkler)	T₁ (mini-sprinkler)	T₂ (micro-sprinkler)
Land preparation	25	25	25	25
Seeds	20	20	20	20
Manures and fertilizers	50	50	50	50
Plant protection	20	20	20	20
Hired human labor	150	120	180	150
Land revenue	25	25	25	25
Total cost per treatment	290	260	320	290
Total cost, Rs./ha	181,250	162,500	200,000	181,250

The total cost of production of cucumber for both the slopes under J-master mini sprinkler and Aqua-master micro-sprinkler irrigation system was Rs. 181,250, Rs. 162,500, Rs. 200,000, and Rs. 181,250, respectively.

2.4.5.1 BENEFIT–COST RATIO OF CUCUMBER PRODUCTION

The economics of cucumber production for the two plots under two irrigation systems was estimated considering total yield per plot, selling price of cucumber, cultivation cost, gross rate of return, net profit, and benefit–cost ratio (Table 2.10). It was found that the benefit–cost ratio for T_1 and T_2 was 1.96 and 2.00 in plot I; 1.26 and 1.30 in plot II, respectively, which indicates that Aqua-master micro-sprinkler irrigation system is capable of high production of cucumber.

TABLE 2.10   Cost Economics of Cucumber Production.

Parameter (units)	Cost economics of production of cucumber			
	Plot I		Plot II	
	T_1 (J-master sprinkler)	T_2 (Aqua-master sprinkler)	T_1 (J-master sprinkler)	T_2 (Aqua-master sprinkler)
Total yield (kg/plot), A	18.8	17.2	13.6	12.7
Total yield (kg/ha), B	11,812.5	10,812.5	8562.5	8000
Selling price of cucumber (Rs./kg), C	30	30	30	30
Cultivation cost (Rs./ha)	181,250	162,500	200,000	181,250
Gross return (Rs./ha), D	354,375	324,375	256,875	240,000
Net return (Rs./ha), E	173,125	161,875	53,750	55,625
B:C ratio, F = D/E	1.96	2.00	1.26	1.30

2.4.5.2 PAY BACK PERIOD

The payback period for cucumber crop was evaluated considering both irrigation systems (Table 2.11). The life of irrigation systems for T_1 and T_2 was considered as 12 and 15 years, respectively. In addition, there are provisions of Government subsidy of 40% for sprinkler irrigation system for the total setup cost in North Eastern Hilly Region. Therefore, the subsidy amount was considered for the said analysis.

TABLE 2.11 Payback Period for Cucumber under Each Treatment.

Parameter (units)	Cost involved for production of Cucumber			
	T_1 (J-master sprinkler)	T_2 (Aqua-master sprinkler)	T_1 (J-master sprinkler)	T_2 (Aqua-master sprinkler)
	Plot I		Plot II	
Irrigation system (Rs./m²)	340	300	360	320
Irrigation system with 40% subsidy (Rs./ha)	204	180	216	192
Total cost (Rs./ha)	2040,000	1800,000	2160,000	1920,000
Net return with irrigation system (Rs./ha) for cultivation per year	2048,125	1800,625	2309,375	2048,750
Life of irrigation system (years)	12	15	12	15
Payback period (years)	5.4	5.2	8.5	8.2

The payback period for the production of cucumber under T_1 and T_2 was 5.4 and 5.2 years, respectively, for the plot I, while for plot II it was 8.5 and 8.2, respectively. It was observed that the payback period is low, though not significant for Aqua-master micro-sprinkler irrigation followed by the J-master mini sprinkler irrigation.

The results obtained in this experimental study are summarized as follows:

- It was observed that the average land slope of about 20% of plot I and 30% of plot II, respectively, indicates a high susceptibility to erosion of soil and a low moisture holding capacity.
- Soil texture for both the plots I and II was found as sandy clay loam with sand, silt, and clay with percentage of 68.2, 8, 27.3 and 64.31, 8.75, 26.9, respectively, for both the plots.
- The soil moisture at FC for both the plots I and II was found to be maximum at depth 0–15 cm (21.76 and 20.06%), respectively, and minimum at the depth of 45–60 cm (13.95 and 12.43%), respectively, and wilting points for plots I and II was found to be maximum at the depth of 0–15 cm (16.79 and 15.26%), respectively, and minimum at the depth of 45–60 cm (13.95 and 12.43%), correspondingly.
- The pH and EC values of collected soil samples from both the treatment areas was found to be in the range of 4.26–5.82 and 0.02–0.49 mS, respectively. It expresses the acidic characteristic of soil and availability of applied fertilizer around 50% to the plant.

- The N, P, K, and OC for plots I and II before plantation was found 115.5, 1.26 mg, 22.5 mg, 0.60%; 112, 1.32, 31.8 mg, 0.537%; and after plantation 125, 1.725, 56 mg, 0.55%; 127.75, 1.15, 53.5 mg, 0.65%, respectively, per unit kilogram of soil sample.
- The discharge, size of laterals and mains, and operating pressure for mini- and micro-sprinkler was found to 7.67 and 10.10 L/h; 16 and 50 mm; 314 and 271 kPa, respectively.
- The growth and yield parameter analysis of cucumber plant grown in the two plots revealed that plants irrigated with Aqua-master micro-sprinkler irrigation system (T_2) provides comparatively better vegetative growth (plant vine length being 51.09%) and yield (number of fruits per treatment being 16.67% and total yield being 0.26%) more than plants irrigated with J-master mini sprinkler irrigation system (T_1).
- The benefit–cost ratio in the plot I for T_1 (mini sprinkler) and T_2 (micro-sprinkler) is 1.33 and 1.51 and plot II for T_1 (mini sprinkler) and T_2 (micro-sprinkler) is 1.25 and 1.58, respectively, which indicates that micro-sprinkler irrigation system is obtainable for the better production of cucumber for a large area.
- The benefit–cost ratio for T_1 and T_2 is 1.96 and 2.00 in plot I; 1.26 and 1.30 in plot II respectively, which indicates that Aqua-master micro-sprinkler irrigation system is feasible to the better production of cucumber.
- The payback period for the production of cucumber under T_1 and T_2 were found 5.4 and 5.2 years, respectively, for the plot I, while for plot II it was found 8.5 and 8.2, respectively. It was observed that the payback period is less though not significantly in the case of the Aqua-master micro-sprinkler irrigation system followed by the J-master mini sprinkler irrigation system. It also reveals that the investments incurred for the setting up of both sprinkler irrigation system could be achieved within the life of protective structures which shows the feasibility and more significant of the system.

2.5 SUMMARY

Plants irrigated with Aqua-master micro-sprinkler irrigation system (T_2) provided comparatively higher values of growth parameters (plant vine length being 51.09%) and yield (number of fruits per treatment being 16.67% and total yield being 0.26%) than plants irrigated with J-master mini sprinkler irrigation system (T_1). Net return was highest for J-master mini sprinkler system (Rs. 2048,125/ha) compared to the Aqua-master micro-sprinkler irrigated system (Rs. 1800,625/ha); however, the payback period for micro-sprinkler was slightly lower compared to mini sprinkler. Therefore, it is suggested to grow cucumber crop with micro-sprinkler irrigation to increase the net annual income in North Eastern Hilly terrains of India.

KEYWORDS

- *Cucumis sativus* L.
- economic analysis
- hilly terrain
- mini sprinkler irrigation

REFERENCES

1. Adetula, O.; Denton, L. Performance of Vegetative and Yield Accessions of Cucumber (*Cucumis sativus* L.). In *Proceedings of the 21st Annual Conference*, 10–13th November; *Horticultural Society of Nigeria (HORTSON)*, 2003; p 200.
2. Aksoy, H.; Kavvas, M. L. A Review of Hillslope and Watershed Scale Erosion and Sediment Transport Models. *Catena* **2005,** *64*, 247–271.
3. Andersen, E.; Baldock, D.; Bennett, H.; Beaufoy, G. *Developing a High Nature Value Farming Area Indicator*. Internal Report for the European Environment Agency, IEEP: Copenhagen, 2003; pp 1–75.
4. Bakker, J. C.; Sonneveld, C. Calcium Deficiency of Glasshouse Cucumber as Affected by Environmental Humidity and Mineral Nutrition. *J. Horticult. Sci.* **1988,** *63* (2), 241–246.
5. Bartosova, M.; Budey, S. Global Challenges for Sustainable Agriculture and Rural Development in Slovakia. *J. Centr. Eur. Agric.* **2013,** *14* (3), 263–278.
6. Berke, T. G.; Black, L. L.; Green, S. K.; Morrish, R. A. *Suggested Cultural Practices for Chilli Pepper*. International Co-operators Guide; AVRDC Publication: Virginia, 1999; p 483.
7. Bhatia, R.; Falkenmark, M. Water Resource Policies and the Urban Poor: Innovative Approaches and Policy Imperatives. In *Proceedings of the International Conference on Water and the Environment: Background Paper for the Working Group on Water and Sustainable Development*, Dublin, 1992; pp 34–35.
8. Bottcher, K.; Eliasson, A.; Jones, R.; Le Bas, C. *Guidelines for Application of Common Criteria to Identify Agricultural Areas with Natural Handicaps*. European Communities; EUR 23795 EN, 2009; pp 1–25.

9. Calzadilla, A.; Rehdanz, K.; Tol, R. S. J. The Economic Impact of More Sustainable Water Use in Agriculture: A Computable General Equilibrium Analysis. *J. Hydrol.* **2010,** *384*, 292–305.

10. Cantliffe, D. J. Alteration of Sex Expression in Cucumber Due to Changes in Temperature, Light Intensity, and Photo Period. *J. Am. Soc. Horticult. Sci.* **1981,** *106* (2), 133–136.

11. Chakraborty, H.; Sethi, L. N.; Lyngdoh, J. Spatio-temporal Rainfall Analysis for Crop Planning in Barak Valley of North East of India. *Silchar* **2014,** *2014*, 1–11.

12. Doll, P.; Siebert, S. Global Modeling of Irrigation Water Requirements. *Water Res. Res.* **2002,** *38*, 1037–1047.

13. Eifediyi, E. K.; Remison, S. U. Growth and Yield of Cucumber (*Cucumis sativus L.*) as Influenced by Farm Yard Manure and Inorganic Fertilizer. *J. Plant Breed. Crop Sci.* **2010,** *7*, 216–220.

14. Eliasson, A.; Jones, R. J. A.; Nachtergaele, F. Common Criteria for the Redefinition of Intermediate Less Favored Areas in the European Union. *Environ. Sci. Policy* **2010,** *13*, 766–777.

15. Fraiture, de C.; Wichelns, D. Satisfying Future Water Demands for Agriculture. *Agric. Water Manage* **2010,** *97*, 502–511.

16. Gobin, A.; Jones, R.; Kirkby, M.; Campling, P. Indicators for Pan-European Assessment and Monitoring of Soil Erosion by Water. *Environ. Sci. Policy* **2004,** *7*, 25–38.

17. Gogoi, M.; Borah, D. *Baseline Data on Area, Production and Productivity of Horticulture Crops in North East and Himalayan States—A Study in Assam, Jorhat*; Assam Agricultural University, 2013; pp 16–67.

18. Grieve, M. Cucumber, **2004**; [Online]. http://www.botanicals.com/botanical/mgmh/c/cucumber123.html (accessed Dec 30, 2019).

19. Howell, T. A.; Cuence, R. H.; Solomon, K. H. Crop Yield Response. In *Management of Farm Irrigation Systems*; Hoffman, G. J., Ed.; ASAE: St. Joseph, 1990; pp 312–318.

20. Jarasiunas, G. Assessment of the Agricultural Land under Steep Slope in Lithuania. *J. Centr. Eur. Agric.* **2016,** *17*, 176–187.

21. Li, J.; Rao, M. Field Evaluation of Crop Yield as Affected by Non-uniformity of Sprinkler-applied Water and Fertilizers. *Agric. Water Manage* **2003,** *59*, 1–13.

22. Liebig, H. P. Physiological and Economical Aspects of Cucumber Crop Density. *Acta Horticult., The Hague* **1981,** *118*, 149–164.

23. Mandave, V. R.; Jadhav, S. B. Performance Evaluation of Portable Mini-sprinkler Irrigation System. *Int. J. Innov. Res. Sci. Eng. Technol.* **2014,** *2014*, 177–184.

24. Marcelis, L. F. M.; Baan Hofman-Eijer, L. R. Effect of Temperature on the Growth of Individual Cucumber Fruits. *Physiologia Plantarum, Kobenhavn* **1993,** *87* (3), 321–328.

25. Medany, M. A.; Wadid, M. M.; Abou-Hadid, A. F. Cucumber Fruit Growth Rate in Relation to Climate. *Acta Horticulturae, The Hague* **1999,** *486*, 107–111.

26. Mitchell, R. D. J.; Harrison, R.; Russell, K. J.; Wess, J. The Effect of Crop Residue Incorporation Date on Soil Inorganic Nitrogen, Nitrate Leaching and Nitrogen Mineralization. *Biol. Fertil. Soils* **2000,** *32*, 294–301.

27. Morgan, R. P. C. *Soil Erosion and Conservation*, 3rd ed.; Blackwell Publishing: Oxford, **2005;** pp 681–687.

28. Namara, R. E.; Upadhyay, B.; Nagar, R. K. *Adoption and Impacts of Micro Irrigation Technologies: Empirical Results from Selected Localities of Maharashtra and Gujarat States of India*. Research Report 93; Colombo, Sri Lanka: International Water Management Institute, **2005;** p 62.

29. Nunn, S. Herbs, **2004**; http://cronescottage2002.tripod.com/ thecottaeaugustmabon2002/ id10.html (accessed Dec 30, 2019).

30. Okonmah, L. U. Effects of Different Types of Staking and Their Cost Effectiveness on the Growth, Yield and Yield Components of Cucumber *(Cumumis sativa* L.). *Int. J. Agric. Sci.* **2011,** *1* (5), 290–295.

31. Orshoven, J. V.; Terres, J. M.; Eliasson, A. *Common Biophysical Criteria to Define Natural Constraints for Agriculture in Europe.* Technical Report, EUR 23412; Joint Research Centre Scientific: Paris, **2008**; pp 1–64.

32. Palanisami, K.; Mohan, K.; Kakumanu, K. R.; Raman, S. Spread and Economics of Micro Irrigation in India: Evidence from Nine States. *Econ. Polit. Wkly* **2011,** *46* (26/27), 81–86.

33. Perry, C.; Steduto, P.; Allen, G. R.; Burt, C. M. Increasing Productivity in Irrigated Agriculture: Agronomic Constraints and Hydrological Realities. *Agric. Water Manage* **2009,** *96*, 1517–1524.

34. Peyvast, Gh.; Olfati, J-Ali; Madeni, S.; Samizadeh, H. Vermi-compost as a Soil Supplement to Improve Growth and Yield of Parsley. *Int. J. Veg. Sci.* **2008,** *14* (1), 82–92.

35. Postel, S. *Pillar of Sand: Can the Irrigation Miracle Last?* : Norton W. W. & Company: New York, 1999; p 312.

36. Saker, N. T.; Capel, P. D. *Environmental Factors That Influence the Location of Crop Agriculture in the Conterminous United States.* U.S. Geological Survey Scientific Investigations Report 5108; U.S. Geological Survey: Washington, DC, 2011; p 72.

37. Saif, U.; Maqsood, M.; Farooq, M.; Hussain, S.; Habib, A. Effect of Planting Patterns and Different Irrigation Levels on Yield and Yield Component of Maize (*Zea mays, L.*). *Int. J. Agric. Biol.* **2003,** *1*, 64–66.

38. Sethi, L. N.; Kumar, N.; Pegu, D. Inter-annual and Spatial Rainfall Analysis for Environmental Restoration in Barak Valley of Assam. *3rd International Conference on "Innovative Approach in Applied Physical, Mathematical/Statistical, Chemical Sciences and Emerging Energy Technology for Sustainable Development"*, 2014; pp 41–46; ISBN: 978-93-83083-98-5.

39. Singh, S. *Taming the Waters: The Political Economy of Large Dam in India*; Oxford University Press: New Delhi, 1997; pp 241–264.

40. Snyder, R. L.; Melo-Abreu, J. P. *Frost Protection: Fundamentals, Practice and Economics*; Food and Agriculture Organization (FAO) of the United Nations: Rome, 2005; p 251.

41. Solaimani, K.; Modallaldoust, S.; Lotfi, S. Investigation of Land Use Changes on Soil Erosion Process Using Geographical Information System. *Int. J. Environ. Sci. Technol.* **2009,** *6* (3), 415–424.

42. Tolk, J. A.; Howell, T. A.; Steiner, J. L.; Krieg, D. R. Role of Transpiration Suppression by Evaporation of Intercepted Water in Improving Irrigation Efficiency. *Irrigation Sci.* **1995,** *16*, 89–95.

43. Vimala, P.; Ting, C. C.; Salbiah, H.; Ibrahim, B.; Ismail, L. Biomass Production and Nutrient Yields of Four Green Manures and Their Effects on the Yield of Cucumber. *J. Tropic. Agric. Food Sci.* **1991,** *27*, 47–55.

44. Weinberger, K.; Thomas A. L. *Diversification into Horticulture and Poverty Reduction: A Research Agenda*; World Bank: Washington, DC, 2007; pp 1464–1480.

45. Yang, X.; Chen, F.; Gong, F.; Song, D. Physiological and Ecological Characteristics of Winter Wheat under Sprinkler Irrigation Condition. *Trans. Chinese Soc. Agric. Eng.,* **2002,** *16* (3), 35–37.

CHAPTER 3

PERFORMANCE OF SELECTED INDIGENOUS CROPS UNDER DRIP IRRIGATION IN THE NORTH-EAST REGION OF INDIA

PANKAJ BARUA

ABSTRACT

Among indigenous horticultural crops, *Citrus limon* L. *Burmf* (Assam Lemon), *Citrus reticulata* L. *Blanco* (Khasi Mandarin), and *Capsicum chinense* (Bhut Jolokia) have commercial potential in eight states of North East Region of India. One common issue associated with commercial growing of these crops is irrigation management. Based on traditional drip irrigation technique using split bamboo, modern techniques were tested for these three crops for effects of water and nutrient levels. All three crops responded well to microirrigation. For Assam Lemon, drip irrigation at 0.8 of water requirement (WR) based on pan evaporation (PE) without black plastic mulch was best treatment for maximum return on investment than the rain-fed conditions. The fertigation through same system resulted in increase in B:C ratio from 3.17 to 4.17 at 20% reduced rate of fertilizer. As an organic alternative, vermiwash was tested for fertigation of Assam Lemon. Despite better growth and yield, there was a negative return on investment because of high cost of vermiwash. Drip irrigation in *C. reticulata* L. *Blanco* (Khasi Mandarin) also resulted in better plant growth. For drip-irrigated *C. chinense* (Bhut Jolokia) under plastic mulching, maximum yield and benefit was obtained when irrigation rate was 1.20 times of WR based on PE.

3.1 INTRODUCTION

Eight states (viz., Arunachal Pradesh, Assam, Manipur, Meghalaya, Mizoram, Nagaland, Sikkim, and Tripura) are together called the North East Region

(NRE) of India. This region is characterized by high hills, narrow river valleys, high rainfall, and moderate climatic conditions. Agro ecology of the region is favorable to grow a wide range of high-value horticultural crops, some of which are quite unique and local to this region. However, productivity of these horticultural crops is very low due to very high rainfall during April through September. Rest of the months are virtually dry with very little or no rainfall. Crops often suffer from moisture stress resulting in yield loss.

Evaporation is comparatively low and it varies from 6 cm in December to 20 cm in May.[4] Therefore, crop water demand is also low. Although agriculture may sustain as sustenance dryland farming, for commercial horticulture, irrigation is essential. High initial investment for head works and undulating topography makes surface irrigation unattractive to farmers. Therefore, drip irrigation has potential to increase crop yield.

A drip irrigation system using bamboo for conveyance has been in use as traditional irrigation technique for horticultural crop in Meghalaya.[7] Scientific work on testing of efficacy of modern drip irrigation technology on crops in North East India was initiated after establishment of Plasticulture Development Center (PDC) at Assam Agricultural University (AAU), Jorhat, in 1989.

First drip irrigation experiment was on *Citrus limon* L. *Burmf* (Assam Lemon), *Citrus reticulata* L. *Blanco* (Khasi Mandarin), *Ananas comosus* (pineapple), and *Brassica oleracea* var. italica (broccoli).[6] Later, research studies were conducted on *Psidium guajava* (guava), banana (cv. Bor Jahaji), *Areca catechu* (arecanut), *Lycopersicon esculentum* Mill. (tomato), *Capsicum annuum* var. grossum (capsicum), *Capsicum chinense* (Bhut Jolokia), chrysanthemum, gerbera, and anthurium. Among the horticultural crops, Assam Lemon, Khasi Mandarin, and Bhut Jolokia are native to this region.

This chapter focuses on use of drip irrigation on three indigenous high-value horticultural crops in North East India. The research works reviewed in this chapter are either carried out or assisted by the author of this chapter.

3.2 EFFICACY OF DRIP IRRIGATION

3.2.1 ASSAM LEMON

Assam Lemon (*C. limon* L. *Burmf*) is an important commercial horticultural crop in North East India. Assam Lemon is grown widely in Assam and occupies an area of 13,000 ha with a productivity of 7 t/ha.[1,2] The fruit lemons are normally consumed raw. The recession of monsoon and subsequent dry months significantly deplete soil water during growing season of the crop. Production is therefore not up to the expected mark.

3.2.1.1   EFFECT OF DRIP IRRIGATION AND PLASTIC MULCH ON ASSAM LEMON

This study was conducted at AAU—Jorhat under Government of India–funded project on "Plasticulture Development Center (PDC)" to study the response of Assam Lemon under drip irrigation to different levels of irrigation with or without black plastic mulch and to investigate the economic viability.[5,6] The study included three different levels of drip irrigation equivalent to 1.0, 0.8, and 0.6 times of water requirement (WR) based on pan evaporation (PE) with and without black plastic mulch. Observations on plant growth and yield were taken. Since the plants are traditionally grown as rained crop, it was taken as the control.

Two years old Assam Lemon plants at 3×3 m spacing were subjected to the treatments for three consecutive years starting in 1998–1999. Observations on plant height, canopy diameter, stem girth, and yield were recorded. Analysis of year-wise and pooled data showed that only drip irrigation and not black plastic mulching influenced the observed growth and yield parameters (Table 3.1). Drip irrigation at "$0.8 \times WR$" without black plastic mulch was best treatment for maximum return on investment at B:C ratio of 3.17 with average water use of 51.37 mm/year and increase in yield of 153% than the rain-fed condition. The net income (Rs./ha) of 74,336.00, 70,016.00, 69,992.00, and 69,728.00 were obtained with plastic mulch along with drip irrigation treatments at 1.0, 0.8, and 0.6 times of WR based on PE, respectively. However, higher return on investment as revealed by ratio of benefit to cost of 3.03, 3.17, and 3.16 could be obtained from drip irrigation without plastic mulch treatments at 1.0, 0.8, and 0.6 times of WR based on PE, respectively. This may be due to high cost of plastic mulch. The net income and benefit to cost ratio of rain-fed plants were much lower irrespective of the black plastic mulch.

3.2.1.2   EFFECT OF DRIP FERTIGATION AND PLASTIC MULCH ON ASSAM LEMON

Since Assam Lemon plants are evergreen in nature, they require adequate water and nutrients throughout the year. Efficacy of drip irrigation and crop

TABLE 3.1 Effects of Irrigation Regimes and Mulching on Yield and Yield Parameters of Drip-irrigated Assam Lemon.

Parameters (year wise and pooled [mean])	A. Moisture regime					B. Mulching			Interaction (CD = 0.5) 1 = A × B 2 = A × Year 3 = B × Year 4 = A × B × Year
	Drip			Rain-fed	CD = 0.5	Nonmulch (NM)	Plastic mulch (PM)	CD = 0.5	
	1.0 WR	0.8 WR	0.6 WR						
Plant height (m)									
1998–1999	1.65	1.59	1.62	1.10	0.08	1.49	1.49	NS	NS
1999–2000	2.03	1.88	1.98	1.39	0.10	1.80	1.84	NS	NS
2000–2001	2.15	1.97	2.08	1.39	0.08	1.89	1.91	NS	NS
Mean	2.04	1.88	1.98	1.35	0.59	1.80	1.82	NS	NS
Canopy diameter (m)									
1998–1999	2.59	2.47	2.47	1.86	0.14	2.36	2.35	NS	1
1999–2000	2.85	2.76	2.70	2.22	0.11	2.60	2.67	NS	1
2000–2001	2.71	2.56	2.59	1.98	0.11	2.41	2.51	NS	NS
Mean	2.75	2.62	2.62	2.07	0.07	2.48	2.55	0.05	2,3
Stem girth (cm)									
1998–1999	5.57	5.52	5.77	2.92	0.48	4.98	4.93	NS	NS
1999–2000	16.31	17.77	17.52	11.10	2.17	15.49	15.86	NS	NS
2000–2001	30.28	28.91	28.54	19.28	2.41	26.34	27.16	NS	NS
Mean	21.39	21.33	21.09	13.82	1.42	19.14	19.67	NS	NS
Yield (number of fruits per plant)									
1998–1999	109.50	102.5	104	37	9.53	83.50	93	NS	NS
1999–2000	107	108	92	37.50	15.02	84.5	88	NS	NS
2000–2001	150	147.3	143.17	74	54.13	125.25	132	NS	NS
Pooled	122.17	119.4	113.05	49.5	17.66	97.75	104.33	4.81	NS

NS, not significant; WR, water requirement.

Source: Adapted from Refs. [1, 2, 3].

WR was standardized by the study reported in Section 3.2.2.1 of this chapter. Objective of this work was to standardize fertigation scheduling for supplying N, P, and K to Assam Lemon plants. To achieve the objective of the study, field experiments were conducted at experimental farm of AAU in Jorhat—India for 3 years starting in 2010. The age of the plantation was 4 years. This work was conducted under Government of India–funded project on "Horticulture Mission for North East India (HMNEI)."

Four fertilizer levels (recommended fertilizer dose [RFD], 20% above RDF and 20% below RDF fertigated) in combination with 50 μm thick black plastic mulching) were tested. Traditional practice of banded application of fertilizer with RDF was used as the control. The results showed 8.47–22.05% increase in yield due to fertigation (Table 3.2). Fertilizer level of 20% above RDF with black plastic mulch showed highest yield and highest net seasonal income of Rs. 231,644.00/ha. Overall black plastic mulch resulted in 14.56–20.53% higher yield than the nonmulched treatment. However, 20% lower RDF without mulch resulted in highest B:C (benefit:cost) ratio of 4.17. Observations on fruit quality, individual fruit weight, fruit volume, and juice content were recorded and analyzed. Fertigation with black plastic mulching significantly enhanced the fruit quality. The experiment concluded that fertigation in combination with black plastic mulching resulted in fertilizer saving and in increase of productivity and profit.

3.2.1.3   ORGANIC FERTIGATION THROUGH VERMIWASH

Considering government's vision of converting NRE of India as organic hub of the country, possibility of organic fertigation was explored by testing efficacy of vermiwash at AAU, Jorhat for 2 years during 2011–2014. Vermiwash is a byproduct of vermi-composting process. Vermiwash is available commercially in the liquid form and was fertigated through drip irrigation in this study. It is rich in nutrients, amino acids, hormones, and useful microbes. Seven treatments for Vermiwash fertigation were:

- Control: Recommended dose of nitrogen (RDN) through soil application;
- RDN and 20% above RDN applied using conventional chemical fertigation;
- 50% below RDN;
- 25% below RDN;
- RDN; and
- 20% above RDN.

TABLE 3.2　Effects of Fertigation and Plastic Mulching on Fruit Yield and Fruit Quality during 2010–2012: Assam Lemon.

Fruit quality	Mulching	Fertilizer level				Mean
		Fertigation			Soil application	
		20% Above RDF	RDF	20% Below RDF	Control, 100% RDF	
Number of fruits per plant	Plastic mulching	123.63	112	102	95.47	108.3
	No mulch	80.77	79.4	72.47	59.63	73.1
	Mean	102.2	95.7	87.23	77.55	–
	$CD_{0.05}$ A. Effect of fertilizer application level					1.9
	B. Effect of mulching					2.7
	A×B					3.8
Fruit yield (t/ha)	Plastic mulching	14.89	12.66	11.07	10.01	10.0
	No mulch	8.06	7.63	6.40	7.74	7.7
	Mean	11.47	10.15	8.73	8.87	–
	$CD_{0.05}$ A. Effect of fertilizer application level					3.1
	B. Effect of mulching					4.4
	A×B					6.3
Fruit weight (g)	Plastic mulching	108.43	101.67	97.67	94.40	100.5
	No mulch	89.83	86.67	79.53	76.07	83.0
	Mean	99.13	94.12	88.60	85.23	–
	$CD_{0.05}$ A. Effect of fertilizer application level					1.7
	B. Effect of mulching					2.3
	A×B					3.3

TABLE 3.2 (Continued)

Fruit quality	Mulching	Fertilizer level					Mean
		Fertigation			Soil application		
		20% Above RDF	RDF	20% Below RDF	Control, 100% RDF		
Fruit volume (cm³)	Plastic mulching	105.50	99.03	95.30	92.07		97.9
	No mulch	86.87	83.70	76.50	75.07		80.5
	Mean	96.18	91.37	85.90	83.57		–
	$CD_{0.05}$	A. Effect of fertilizer application level					1.7
		B. Effect of mulching					2.4
		A×B					3.4
Juice content (%)	Plastic mulching	42.6	43.0	42.8	40.6		42.3
	No mulch	42.3	39.8	41.5	41.4		41.3
	Mean	42.5	41.5	42.2	41.1		–
	$CD_{0.05}$	A. Effect of fertilizer application level					0.006
		B. Effect of mulching					0.009
		A×B					0.013

The results for the first two years of trials (Table 3.3) revealed that different treatments failed to influence the yield of Assam Lemon plants significantly. However, treatments became significant during third and fourth year of the experimentation. It was interesting to observe that even 50% below RDN applied through vermiwash using low cost resulted in better or at par yield compared to that of RDN through soil application (control). Vermiwash application through low-cost drip irrespective of levels resulted in better yield than the conventional drip fertigation. The yield of Assam Lemon plants with 50% below RDN applied through vermiwash using low-cost drip was at par with RDN application through fertigation during 2012–2013, 2013–2014 and in pooled data. The 20% above RDN applied through vermiwash using low-cost drip resulted in the highest yield during all the years of experimentation and in pooled data. However, the net seasonal income and benefit cost ratio (BCR) was found to be highest (Rs. 254,500/ha and 3.85) in case of 20% above RDN applied using conventional chemical fertigation. There was negative BCR in the treatments with vermiwash fertigation because of high cost of vermiwash (@ Rs. 2/L). This study was conducted under All India Coordinated Research Project on Irrigation Water Management (AICRPIWM) of Indian Council of Agricultural Research (ICAR).

3.2.1.4   RAIN WATER HARVESTING AND DRIP IRRIGATION

NRE of India receives very high rainfall and its distribution is not uniform. High rainfall occurs during April to September with very little or no rain in other months. Harvesting of rain water and subsequent use for domestic purpose is common. Since water harvesting involves extra cost, use of the harvested water judiciously through drip irrigation is a possibility. A high-value low-duty crop should be chosen. This approach was tested at the experimental farm of AAU—Jorhat during 2009–2011. A pond was dug and was lined with plastic film to reduce seepage losses. During the rainy season, rain water was harvested in a pond.

A plantation of 2 years old Assam Lemon crop planted at 3 × 3 m spacing was drip irrigated using the harvested water in the pond during the dry months starting in November. The 50-µm thick black plastic mulch was used to conserve water. Three irrigation levels were tested, such as 1.0, 0.8, and 0.6 times of PE based on USDA Class-A pan. Fruit yield and plant attributes were observed and analyzed. Results revealed that only

TABLE 3.3 Effect of Vermiwash (Liquid Fertilizer) on Number of Fruits per Plant for Assam Lemon.

Description	Treatment	Yield (no. of fruits per plant)				
		2010–2011	2011–2012	2012–2013	2013–2014	Pooled
Rain-fed + soil application (RDN)	T_1	20.00	32.75	74.00	172.50	74.81
Fertigation (20% above RDN) conventional drip	T_2	18.25	33.50	116.00	237.25	101.25
Fertigation (RDN) conventional drip	T_3	21.50	31.25	87.50	223.00	80.81
Vermiwash (20% above RDN) through low cost drip	T_4	24.0	34.00	118.25	238.75	103.75
Vermiwash (RDN) through low cost drip	T_5	24.75	30.50	98.00	221.00	93.56
Vermiwash (25% below RDN) through low cost drip	T_6	24.5	34.50	84.00	209.50	88.12
Vermiwash (50% below RDN) through low cost drip	T_7	21.25	32.50	70.50	195.25	79.87
CD at P = 5%		NS	NS	16.07	39.70	10.7
CV		31.22	21.20	11.69	12.50	16.86

TABLE 3.4 Benefit Cost Ratio (BCR) for Different Treatments.

Item	Treatments							
	Drip (1.0 PE) + no mulch	Drip (0.8 PE) + no mulch	Drip (0.6 PE) + no mulch	Drip (1.0 PE) + plastic mulch	Drip (0.8 PE) + plastic mulch	Drip (0.6 PE) + plastic mulch	Rain-fed + no mulch	Rain-fed + plastic mulch
No. of fruits (,000/ha)	122.08	127.77	127.41	145.67	140.24	121.11	50.48	64.56
Average water applied (mm/year)	64.21	51.37	38.49	64.21	51.37	38.49	–	–
Income for lemon (Rs.)	65,464	70,016	69,728	74,336	69,992	54,688	15,384	16,648
BCR	2.10	2.22	2.21	2.33	1.82	1.48	0.81	0.64

drip irrigation levels influenced the observed parameters and that this technology increased the crop productivity and profit. The BCR (benefit-to-cost ratio) was highest for water application of 1.0 times PE with black plastic mulching. This work was conducted under Government of India–funded project on "HMNEI".

3.2.2 KHASI MANDARIN

Khasi Mandarin (*C. reticulata* L. *Blanco*) is an indigenous crop in North-East region of India. It is one of the most popular fruits due to its high palatability and richest source of minerals and vitamins. Khasi Mandarin is grown in almost all states of the NRE. In Assam, the area coverage is about 6000 ha. The crop is mostly grown as rain-fed crop, and therefore, growth and yield are affected due to moisture stress during winter.

3.2.2.1 EFFECT OF DRIP IRRIGATION ON KHASI MANDARIN

This work was conducted at AAU—Jorhat under Government of India–funded project on PDC with the objective of studying the response of Khasi Mandarin to different levels of drip irrigation with or without black plastic mulch and to investigate the economic viability. Three different levels of drip irrigation equivalent to 1.0, 0.8, and 0.6 times of WR based on PE with and without black plastic mulch were tested. Observations on plant growth parameters were recorded.

The results revealed that plant height, canopy area, and stem girth were significantly superior in all drip-irrigated treatments compared to the rain-fed treatment (Table 3.5). Highest values of physiological parameters (plant height, canopy area, and stem girth) were recorded when the plants were drip irrigated with 101.08, 91.48, 45.34, 99.20, and 65.69 L of water during November, December, January, February, and March, respectively; and the plots were mulched with black plastic film of 50 μm thickness.

3.2.3 BHUT JOLOKIA

Bhut Jolokia (*C. chinense Jacq.*) is also known as Borbih Jolokia, Naga Jolokia, Nagahari, Naga Morich, and Raja Mirchi, and is extensively cultivated in North-East Region of India (especially in the states of Assam,

Nagaland, and Manipur). It has been reported as the "hottest chilli on earth" with Scoville Heat Unit of 855,000.[5,6] This has created lot of interest in this crop and has opened up avenues for commercialization. Although no clear data is available on the area under this crop, it is steadily increasing during the last few years. The crop is cultivated as dryland crop.

TABLE 3.5   Physiological Parameters, Pooled Data (October 1998 through Feb. 2000): Khasi Mandarin.

Treatment	Description	Plant height (m)	Stem girth (m)	Canopy area (m²)
T_1	Drip (1.0 PE) + no mulch	1.63	0.096	0.618
T_2	Drip (0.8 PE) + no mulch	1.82	0.103	0.769
T_3	Drip (0.6 PE) + no mulch	1.53	0.093	0.605
T_4	Drip (1.0 PE) + plastic mulch	1.71	0.099	0.770
T_5	Drip (0.8 PE) + plastic mulch	1.52	0.089	0.610
T_6	Drip (0.6 PE) + plastic mulch	1.56	0.084	0.687
T_7	Rain-fed + no mulch	1.24	0.073	0.518
CD at 5%				
Treatment		0.117	0.007	0.065
Season		0.103	0.007	0.074
Treat × season		0.310	0.019	0.196

TABLE 3.6   Effects of Drip Irrigation and Plastic Mulching on Yield of *Bhut Jolokia.*

Treatments		No. fruits pWR plant	Yield (kg/plant)	Yield (t/ha)
120% WR + PM	T_1	138.25	1.51	15.10
100% WR + PM	T_2	138.22	1.50	15.00
80% WR + PM	T_3	133.22	1.24	12.40
60% WR + PM	T_4	52.43	0.41	4.10
120% WR	T_5	128.16	1.14	11.40
100% WR	T_6	134.52	1.17	11.70
80% WR	T_7	101.53	0.81	8.10
60% WR	T_8	41.38	0.30	3.00
	CD	11.6	0.10	
	CV	6.14	5.72	

WR, water requirement based on PE; PM, plastic mulch.

3.2.3.1   EFFECT OF DRIP IRRIGATION ON BHUT JOLOKIA

This study was conducted at AAU—Jorhat under AICRPIWM of ICAR with the objective of studying the response of drip-irrigated Bhut Jolokia to different irrigation levels with or without black plastic mulch and to investigate the economic viability. Four different irrigation levels equivalent to 1.2, 1.0, 0.8, and 0.6 times of WR based on PE with and without black plastic mulch were tested. The results of the study revealed that different irrigation levels and black plastic mulch significantly affected yield and number of branches.

The highest yield (1.50 kg/plant) was observed in plants with highest water application (1.20 times of WR based on PE) and under plastic mulching. The lowest yield (0.66 kg/plant) was observed in plants with lowest water application (0.6 times of WR based on PE).[8,9] Economic analysis revealed highest B:C ratio of 9.10.

3.3   SUMMARY

This chapter includes performance of *C. limon* L. *Burmf* (Assam Lemon), *C. reticulata* L. *Blanco* (Khasi Mandarin), and *C. chinense* (Bhut Jolokia) under drip irrigation and plastic mulching. For all the crops, there was increase in growth parameters and yield compared to rain-fed plants. The results of the experiment on integration of rainwater harvesting and drip irrigation will encourage sustainable use of harvested rain water in hilly areas. Fertigation through drip resulted in increase in B:C ratio in *C. limon* L. *Burmf* (Assam Lemon).

ACKNOWLEDGMENT

This research study was carried out with funding from Department of Agriculture and Cooperation, Ministry of Agriculture, Government of India, through Plasticulture Development Center (PDC) and Horticulture Mission for North East India (HMNE) and Indian Council of Agricultural Research (ICAR) under All India Coordinated Research Project on Irrigation Water Management (AICRPIWM).

KEYWORDS

- **Assam**
- **Bhut Jolokia**
- **drip irrigation**
- **Khasi Mandarin**
- **lemon**
- **plastic mulching**

REFERENCES

1. AICRPIWM. *Annual Report: All India Coordinated Research Project on Irrigation Water Management* (AICRPIWM). Assam Agricultural University (AAU): Jorhat, 2017; pp 45–60.
2. Barua, P.; Hazarika, R. Studies on Fertigation and Soil Application Methods along with Mulching on Yield and Quality of Assam Lemon (*Citrus limon* L. Burmf.). *Indian J. Horticult.* **2014,** *71* (2), 190–196.
3. NERIWALM (North Eastern Regional Institute of Water and Land Management). *Some High Value Horticultural Crops Grown in NER of India,* 2018. http://neriwalm.gov.in/news.html (accessed Aug 26, 2019).
4. NIH (National Institute of Hydrology). *Hydrology and Water Resources Information System for India, Climate, Evaporation,* 2018. http://nihroorkee.gov.in/rbis/india_information/evaporation.htm (accessed Aug 23, 2019).
5. PDC (Plasticulture Development Center). *Preliminary Report, 1988–89.* Plasticulture Development Center (PDC), Jorhat Center, AAU: Jorhat, 1989; pp 1–4.
6. PDC (Plasticulture Development Center). *Annual Report 1999–2000.* Plasticulture Development Center (PDC), Jorhat Center, AAU: Jorhat, 2000; pp 15–35.
7. Sen, P.; Devi, N. L.; Singha, D. Understanding Indigenous Irrigation Systems in North East India. *Eco. Whisper* **2015,** *2* (1), 8–15.
8. Singh, A.; Singh, B. Khasi Mandarin: Its Importance, Problems and Prospects of Cultivation in North-Eastern Himalayan Region. *Int. J. Agric. Environ. Biotechnol.* **2016,** *9* (4), 573–592.
9. Tiwari, K. N. *Micro Irrigation System Design,* 2018. http://ecoursesonline.iasri.res.in/course/view.php?id=546 (accessed Aug 25, 2019).

CHAPTER 4

PERFORMANCE OF POINTED GOURD UNDER POLYETHYLENE MULCHING AND DIFFERENT LEVELS OF FERTIGATION

ABINASH DALAI and P. C. PRADHAN

ABSTRACT

Field experiments were conducted at Odisha University of Agriculture and Technology—Bhubaneswar to evaluate the response of drip fertigation and polyethylene mulching in pointed gourd during 2015–2017. The treatments were T_1, T_2, T_3 corresponding to three fertigation levels, namely, 100, 80, and 60% of recommended dose of fertilizer (RDF); and T_4, T_5, and T_6 corresponding to fertigation levels of 100, 80, and 60% of RDF along with polyethylene mulching. The control treatment (T_7) consisted of basal application of 100% NPK along with drip irrigation (DI). The biometric parameters for pointed gourd under 100% fertigation with mulching recorded an increase of vine length by 53%, stem girth by 16%, leaf area by 33%, and primary branches by 31.6%, respectively, compared to the control. The marketable yield was maximum (15.81 t/ha) in T_4, which was at par with T5 (14.4 t/ha) and there was 26.3% increase in yield over the control. All treatments under drip fertigation with mulching showed 8.5% higher IWUE compared to nonmulched treatments. Fertilizer use efficiency (FUE) was increased with decrease of fertigation level. The maximum FUE of 65.6 kg/kg of NPK was observed for 60% fertigation with mulching. Also, there was inverse relationship between fertigation level and FUE.

4.1   INTRODUCTION

Agricultural sector is a voracious consumer of fresh water. The water is getting pressurized due to demand from different sectors of life. The climate change will lead to water shortage, which may lead to increased frequency of droughts. A large part of India is already experiencing water stress. The temporal and spatial distribution rainfall is not uniform. About 80% of annual rainfall occurs only in four monsoon months from June to September and rest 20% rainfall spreads over 8 months. According to the statistics of Water Resources Department—Government of India, the share of irrigation water was 78% in 2000 AD and it will be reduced to 64% by 2050 due to stiff competition from other sectors, such as industries, domestic, and municipal uses. The normal annual rainfall of Odisha is 1452 mm, which is equivalent to an annual precipitation of about 230.8 BCM (billion cubic meters). Keeping in view of the reduced per capita availability of water and present level of water utilization for irrigation, it is of paramount importance to plan for the high-efficiency irrigation systems. This requires enhancing of irrigated area and production for the growing population on sustainable basis.[2,3]

The drip irrigation (DI) is an advanced irrigation method, which has the potential to meet the nutritional and water requirement of crops during various growth stages thereby increasing the crop productivity and quality of produce. In this method, water is applied slowly and frequently at the root zone catering to the daily needs. Experiment conducted by Tiwari et al. on drip-irrigated okra under plastic mulching showed that irrigation based on 100% irrigation requirement gave the maximum yield of 14.5 t/ha with 72% higher yield compared to furrow irrigation.[20] Another study on broccoli Kumari et al. indicated that the highest yield of 24.46 t/ha was observed with 80% ETc followed by 100% ETc, when irrigation was scheduled on alternate days.[11]

India has achieved 80% self-sufficiency in production capacity of urea and 50% indigenous capacity for phosphate fertilizers. However, 100% potash fertilizers are imported to meet the national need. The benefits of DI are not fully harnessed, because fertigation is not practiced in most cases either due to lack awareness or absence of fertigation scheduling in drip-irrigated vegetables in Odisha. The field study showed the highest yield of drip-irrigated okra (24.91 t/ha) with 100% RDF (recommended dose of fertilizer).[16] In case of guava, research study showed that the interaction effect of 100% ETc and N-fertigation of 120% RDF gave the maximum yield of 21.6 t/ha and water productivity of 17.8 kg/ha mm.[18] It was revealed that DI at 80% water requirement and 60% NPK fertigation (recommended) resulted in fruit yield of 98.93 t/ha, high water productivity of 4.99 t/ha, and fertilizer use efficiency (FUE) of 1.09, 1.83, and 2.74 t/kg of NPK, respectively.[3,8] Plastic mulch is used to reduce soil moisture, weed infestation and to prevent nutrient loss.[7] Field experiment at Jorhat—Assam revealed that 120% RDF with plastic mulch resulted in maximum yield of Assam lemon of 14.88 t/ha.[4]

This chapter focuses on growth, productivity, and input use efficiency of pointed gourd under various levels of fertigation and polyethylene mulching in coastal plain zone of Odisha—India.

4.2   METHODS AND MATERIALS

Experiment was conducted during 2015–2016 and 2016–2017 with growing season from the third week of December to first week of November. The field experiment consisted of design and installation of drip fertigation system, plantation of pointed gourd, irrigation scheduling, fertigation scheduling, field observation, and analysis of data.

4.2.1   EXPERIMENTAL SITE

The experiment was carried out at Precision Farming Development Centre (PFDC), Odisha University of Agriculture and Technology (OUAT), Bhubaneswar that is located between 20°31′15″N latitude and 85°47′24″E longitude (Fig. 4.1). The field capacity and permanent wilting point of soil at the study site were 18.9 and 3%, respectively, on volume basis. The bulk density of soil was 1.52 g/cm^3. The properties of soil are given in Table 4.1.

4.2.2   EXPERIMENTAL DESIGN

The field experiment was conducted using randomized block design. There were seven treatments, such as:

- T_1, T_2, and T_3 relate to fertigation levels of 100, 80, and 60% of RDF.
- T_4, T_5, and T_6 relate to fertigation levels of 100, 80, and 60% of RDF with polyethylene (LDPE) mulching.
- T_7 relates to DI with basal application of 100% NPK (control).

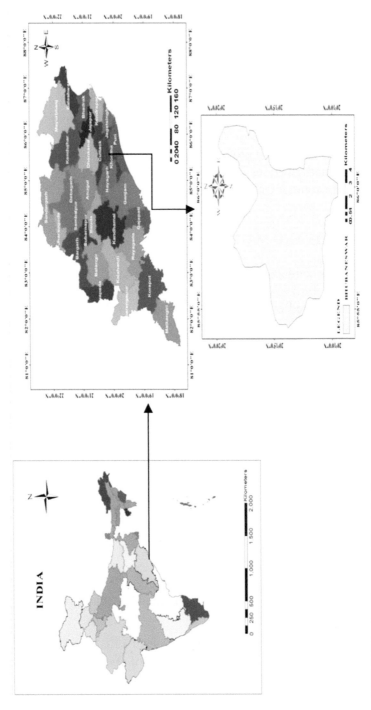

FIGURE 4.1 Location map of the study site.

TABLE 4.1 Soil Properties at Experimental Site.

Parameter	Soil depth		
	0–15 cm	**15–30 cm**	**30–45 cm**
Textural class	**Sandy loam**	**Loamy sand**	**Sandy loam**
pH	4.44	4.48	4.50
EC (dSm^{-1})	0.07	0.07	0.08
OC (%)	0.40	0.37	0.35
Available nitrogen (kg/ha)	164.0	175.0	190.0
	30.0	38.0	36.0
	100.0	78.0	70.0
	24.0	28.0	26.0
Available boron (mg/kg)	0.2	0.3	0.2
Available zinc (mg/kg)	0.36	0.40	0.42

After the field preparation, the farmyard manure was added at 20 t/ha. Then 21 plots of 6 m × 5 m were laid out in the field in three replications. Each plot consisted of five beds of 1m in width. The 16 mm laterals were laid at 1.2 m spacing and online drippers of 4 lph were used at a spacing of 1 m. PVC pipes of 63 and 50 mm diameter were used for main and submain pipes, respectively. The water from the tube well was passed through a hydrocyclone and screen filters. The system was operated at constant pressure of 1.0 kg/cm.

4.2.3 WEATHER PARAMETERS DURING CROP SEASON

4.2.3.1 THE FIRST YEAR, 2015–2016

Monthly rainfall was 264.8 mm during June 2016 followed by 247.8 mm in August 2016. The minimum monthly rainfall was during January 2016. The maximum 24 rainy days were recorded in September 2016 followed by 22 rainy days in August 2016. The maximum mean monthly evaporation was 7.1 mm during May 2016 and the minimum mean monthly value of 3.2 mm during August 2016 during crop growth period (Fig. 4.2a). The maximum mean monthly temperature was 40.8°C during May 2016 and the minimum mean monthly temperature was 15.7°C during January 2016. The minimum relative humidity during morning was 36% in December 2015 and the maximum value was 93% in August and September 2016. However, the minimum relative humidity during afternoon was in April 2016 and maximum value of 80% during September 2016 (Fig. 4.2b).

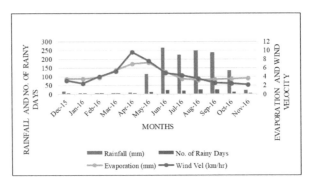

FIGURE 4.2a Rainfall, rainy days, evaporation, and wind velocity during cropping season, 2015–2016.

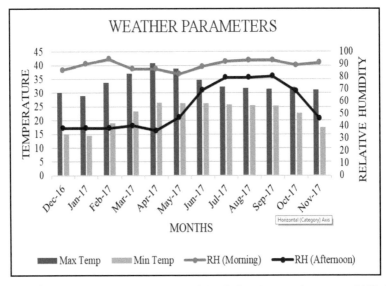

FIGURE 4.2b Temperature and relative humidity during the cropping season, 2015–2016.

4.2.3.2   THE SECOND YEAR, 2016–2017

Maximum rainfall was 473.2 mm during June 2017 followed by 393.7 mm in August 2017. The lower monthly rainfall was recorded during December 2016, January and February 2017. The maximum 20 rainy days were recorded in July 2017 followed by 18 rainy days in August 2017. The maximum mean monthly evaporation was 8.7 mm during May 2017 and minimum mean monthly value was 2.9 mm during July 2017 during crop season (Fig. 4.3a). The maximum mean monthly temperature was 38.8°C during May 2017 and minimum mean monthly temperature was 14.2°C during January 2017. The relative humidity during morning time was the lowest with value of 82% during May 2017 and the highest value was 94% during February 2017. However, relative humidity during afternoon was the lowest value in December 2016 and was the maximum with a value of 78% during July 2017 during the crop season (Fig. 4.3b).

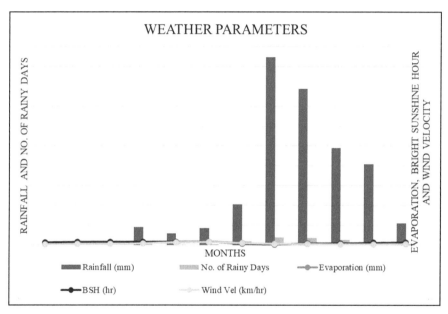

FIGURE 4.3a Rainfall, rainy days, evaporation, and wind velocity during crop season, 2016–2017.

4.2.4 CULTIVATION OF POINTED GOURD

The rooted cuttings of pointed gourd were planted in rows at 120 cm × 100 cm spacings during the third week of December in both years (2015 and 2016) (Fig. 4.4). The recommended package practices for pointed gourd were followed. The plastic mulching was 100-μm bicolor (silver and black) low-density polyethylene film in treatments T_4, T_5, and T_6. DI was used to apply water on alternate days in all treatments.

The fertigation unit was Venturi injector. The water-soluble fertilizers for the experiments are urea (46:0:0), urea phosphate with SOP (18:18:18), and sulfate of potash (0:0:50) based on RDF of 188:60:100 :: N:P:K. The different levels of fertigation were applied at monthly intervals. Crop growth parameters and yield were recorded in all treatments.

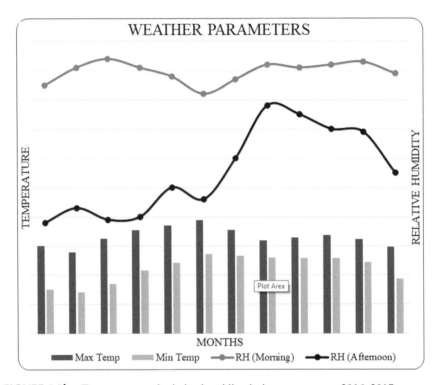

FIGURE 4.3b   Temperature and relative humidity during crop season, 2016–2017.

4.2.5 STATISTICAL ANALYSIS

The biometric observations were yield and input use efficiencies that were analyzed using ANOVA.[12]

4.3 COMPUTATION OF CROP WATER REQUIREMENTS (CWR)

The quantity of water (liters/day) through the DI system to each plant was calculated with the following equation[14]:

$$V = \frac{\left[ET_o \times K_c \times S_l \times S_e \times W_s \right]}{\eta} \tag{4.1}$$

where V is the volume of irrigation water applied (liters/day/plant), ET_o is the reference crop evapotranspiration (mm/day) calculated by FAO-56 Penman–Monteith method,[1] K_c is the crop coefficient, S_l and S_e is the lateral and emitter spacings taken as 1.2, and 0.4 m, respectively, W_s is the percentage wetted area factor, and η is the emission uniformity of the system.

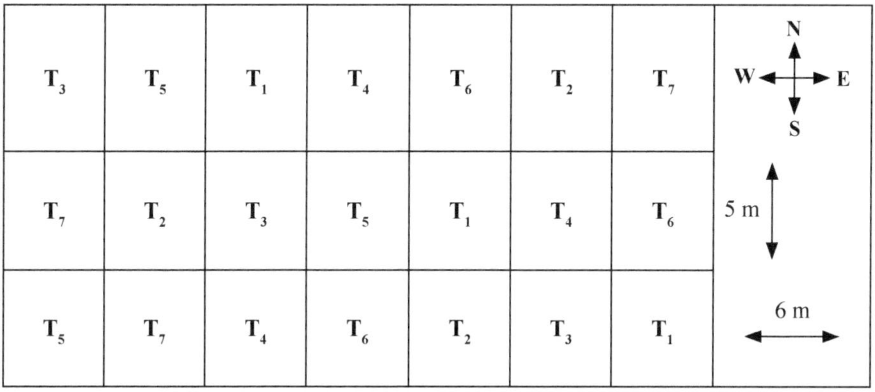

T_3	T_5	T_1	T_4	T_6	T_2	T_7	N W ← ✛ → E S
T_7	T_2	T_3	T_5	T_1	T_4	T_6	5 m
T_5	T_7	T_4	T_6	T_2	T_3	T_1	6 m

FIGURE 4.4   Layout of field experiment (row–row spacing and plant–plant spacing: 5 m × 6 m).

4.4   IRRIGATION EFFICIENCIES

4.4.1   IRRIGATION WATER-USE EFFICIENCY (IWUE)

It refers to the crop yield per irrigation used to produce this yield.[10] The IWUE was determined using the following equation[6]:

$$\text{IWUE} = \left(\frac{E_y}{I}\right) \qquad (4.2)$$

where IWUE is the irrigation water-use efficiency (kg/m³), E_y is the yield (kg/ha), and I is the volume of irrigation (m³).

4.4.2   FERTILIZER USE EFFICIENCY

It refers to the crop yield per kg of fertilizer to produce this yield.[13] Therefore, FUE was calculated as "the fresh fruit weight (kg) per unit weight of fertilizer used (kg)."

4.5   RESULTS AND DISCUSSION

4.5.1   BIOMETRIC OBSERVATIONS

The biometric observations were vine length, stem girth, leaf area, and number of primary branches at fortnight intervals. The observations at 120 DAP (days after planting) were taken into consideration for statistical analysis and are given in Table 4.2.

TABLE 4.2   Growth Parameters of Pointed Gourd for Different Treatments.

Treatment	Vine length	Girth of stem	Leaf area	No. of Primary Branches
	cm		cm^2	No.
T$_1$ (fertigation at 100% RDF without mulch)	397.88	2.78	111.42	7.25
T$_2$ (fertigation at 80% RDF without mulch)	376.30	2.72	102.41	7.03
T$_3$ (fertigation at 60% RDF without mulch)	316.47	2.55	89.48	6.22
T$_4$ (fertigation at 100% RDF with mulch)	405.50	2.91	113.59	7.88
T$_5$ (fertigation at 80% RDF with mulch)	396.28	2.79	111.65	7.48
T$_6$ (fertigation at 60% RDF with mulch)	327.12	2.65	98.78	6.67
T$_7$ (drip irrigation with 100% RDF basal application)	264.12	2.47	85.29	6. 02
SEm(±)	15.22	0.03	0.55	0.20
CD (P = 0.05)	47.42	0.10	1.71	0.62

The treatments under drip fertigation with polyethylene mulching showed better length of vine compared to treatments without mulching. The 100% fertigation through drip with mulching recorded maximum vine length of 4.05 m, which was at par with 80% fertigation with mulching and there was 53% higher growth over the control. Similarly, the maximum value of stem girth was 2.91 cm in T$_4$ and was significantly superior to the other treatments.

The maximum value of leaf area was 113.59 cm^2, which was significantly higher than the other treatments and 33% higher than the control. Plants under mulching treatments indicated 6.85% greater leaf area than the plants without mulching. The number of primary branches was higher in T$_4$ and at par with T$_5$, but significantly higher than other treatments. This improved growth may be attributed to regular and timely supply of required water and nutrients through drip fertigation, which might have enhanced the uptake of the nutrients.[19] The maintenance of favorable soil moisture and easy availability of nutrients in the active root zone was able to increase the growth parameters of pointed gourd under drip fertigation. Similar results were observed in case of tomato at 100% N & K fertigation.[5]

4.5.2 YIELD AND YIELD ATTRIBUTING CHARACTERS

The pooled data of 2 years were analyzed for variance (Table 4.2). All drip fertigation treatments with mulching recorded 16% higher fruit weight over nonmulched treatments. The average maximum fruit weight was 28.3 g in 100% fertigation with mulch (T_4). Similarly, the average fruit length was 10.3% higher values in T_4, T_5, and T_6 compared to nonmulched treatments, and it was 41.5% over the control. The fruit length was 9.25 cm in T_4 but at par with T_5 (8.4 cm). Likewise, average fruit girth for drip-fertigated treatments with mulching was 5.9% higher compared to only drip fertigation treatments. The maximum value of fruit girth was 9.28 cm for T_4, which was at par with T_5 and 19.5% higher values over control.

The maximum fruit yield was 15.8 t/ha in T_4, which was at par with T_5 (14.4 t/ha), but it increased to 26.3% over the control. Also, drip-fertigated treatments with mulching recorded 8.6% higher yield over nonmulched treatments. Similarly, the maximum yield of tomato was reported by researchers in 100% fertigation.[9,15] Furthermore, the highest yield of pointed gourd was under 100% NPK drip fertigation and plastic mulching.[17]

TABLE 4.3 Yield and Yield Attributes of Pointed Gourd for Different Fertigation Levels and Mulching.

Treatment	Fruit weight	Fruit length	Fruit girth	Yield
	g		cm	t/ha
T_1 (fertigation at 100% RDF without mulch)	23.92	8.01	8.52	14.41
T_2 (fertigation at 80% RDF without mulch)	22.95	7.76	8.22	13.97
T_3 (fertigation at 60% RDF without mulch)	19.80	6.87	7.49	13.06
T_4 (fertigation at 100% RDF with mulch)	28.29	9.25	9.28	15.81
T_5 (fertigation at 80% RDF with mulch)	27.25	8.44	8.71	15.49
T_6 (fertigation at 60% RDF with mulch)	21.72	7.28	7.70	13.72
T_7 (DI with 100% soil based application of RDF)	18.10	6.48	7.24	12.51
SEm(±)	0.30	0.16	0.26	0.19
CD (P = 0.05)	0.94	0.48	0.82	0.60

4.5.3   INPUT USE EFFICIENCY

Greater reduction in total water used for the production of pointed gourd was observed in DI over surface irrigation. The maximum value of IWUE was 3.5 kg/m^3 (T_4), which was at par with T_5 (3.4 kg/m^3 of water), but was 26.6% higher than the control. All treatments of drip fertigation with mulch showed 8.5% higher IWUE as compared to nonmulched treatments.

FUE was increased with decrease in fertigation levels. The maximum FUE was 65.6 kg/kg of NPK in the case of 60% fertigation with mulch (T_6) followed by T_3 (62.5 kg/kg of NPK). The results showed that there exists an inverse relationship between fertigation level and FUE (Table 4.4). It is also confirmed that there was saving of fertilizer compared to T_7, where water and nutrients to the plants and the water were delivered to the plants frequently in the root zone.

TABLE 4.4   Input Use Efficiency for Different Treatments.

Treatment	Yield	IWUE	FUE
	t/ha	kg yield per m^3 of water	kg yield per kg of NPK fertilizer
T_1 (fertigation at 100 % RDF without mulch)	14.41	3.20	41.39
T_2 (fertigation at 80 % RDF without mulch)	13.97	3.11	50.26
T_3 (fertigation at 60% RDF without mulch)	13.06	2.90	62.50
T_4 (fertigation at 100% RDF with mulch)	15.81	3.52	45.42
T_5 (fertigation at 80% RDF with mulch)	15.49	3.44	55.74
T_6 (fertigation at 60% RDF with mulch)	13.72	3.05	65.62
T_7 (drip irrigation with 100% RDF in soil application)	12.51	2.78	35.94
SEm (±)	0.19	0.05	0.69
CD (P = 0.05)	0.60	0.14	2.15

4.6   SUMMARY

The growth parameters of pointed gourd in treatment combinations of 100% drip fertigation of NPK with mulching showed an increase of vine length by 53%, stem girth by 16%, leaf area by 33%, and number of primary branches by 31.6% compared to 100% basal application of fertilizer and DI (control). Similarly, maximum yield and IWUE were in case of 100% fertigation with mulching and it was 26, 27% higher, respectively, over the control. However, the FUE was increased with decrease of fertigation level and maximum value was in the case of 60% fertigation with mulching. The mulched treatment showed 8.2% higher FUE compared to nonmulched treatments.

ACKNOWLEDGMENT

Authors are thankful to Ministry of Agriculture & Farmers' Welfare, Govt. of India, New Delhi; and NCPAH, New Delhi for providing the financial support to conduct this research.

KEYWORDS

- fertilizer use efficiency
- irrigation water-use efficiency
- low-density polyethylene
- precision farming

REFERENCES

1. Allen, R. G.; Pereira, L. S.; Raes, D.; Smith, M. *Crop Evapotranspiration: Guide-lines for Computing Crop Water Requirements*. FAO Irrigation and Drainage Paper 56; Food & Agricultural Organization: Rome, Italy, 1998; p 300.
2. Anonymous. *National Water Policy (NWP): Unpublished Report*; Ministry of Water Resources, Govt. of India: New Delhi, 2012; p 13.
3. Badr, M. A.; El-Tohamy, W. A.; Zaghloul, A. M. Yield and Water Use Efficiency of Potato Grown under Different Irrigation and Nitrogen Levels in an Arid Region. *Agric. Water Manage.* **2012,** *110,* 9–15.
4. Barua, P.; Hazarika, R. Studies on Fertigation and Soil Application Methods Along with Mulching on Yield and Quality of Assam Lemon (*Citrus limon* L. Burmf.). *Indian J. Hort.* **2014,** *71* (2), 190–196.
5. Brahma, S.; Phookan, D. B.; Barua, P.; Saikia, L. Effect of Drip Fertigation on Performance of Tomato under Assam Condition. *Indian J. Hort.* **2010,** *67* (1), 56–60.
6. Cetin, O.; Uygan, D. The Effect of Drip Line Spacing, Irrigation Regimes and Planting Geometries of Tomato on Yield, Irrigation Water Use Efficiency and Net Return. *Agric. Water Manage.* **2008,** *95,* 949–958.
7. Fritz, V. A. *Plastic Mulches: Benefits, Types and Sources.* Minnesota High Tunnel Production Manual for Commercial Growers, 2nd ed.; Regents of the University of Minnesota, 2012; p 168; http://hightunnels.cfans.umn.edu/ (accessed Feb 18, 2019).

8. Gupta, A. J.; Chattoo, M. A.; Singh, L. Drip Irrigation and Fertigation Technology for Improved Yield, Quality, Water and Fertilizer Use Efficiency in Hybrid Tomato. *J. Agrisearch* **2014,** *2* (2), 94–99.

9. Hebbar, S. S.; Ramachandrappa, B. K.; Nanjappa, H. V.; Prabhakar, M. Studies on NPK Drip Fertigation in Field Grown Tomato. *Eur. J. Agron.* **2004,** *21,* 117–127.

10. Howell, T. A. Irrigation Scheduling Research and Its Impact on Water Use. In *Proceedings of the International Conference on Evapotranspiration and Irrigation Scheduling*; ASABE: St. Joseph, **1996;** pp 21–33.

11. Kumari, A.; Patel, N.; Mishra, A. K. Response of Drip Irrigated Broccoli (Brassica oleracea var. italica) in Different Irrigation Levels and Frequencies at Field Level. *J. Appl. Nat. Sci.* **2018,** *10* (1), 12–16.

12. Panse, V. G.; Sukhatme, P. V. *Statistical Methods for Agricultural Workers*, Indian Council of Agricultural Research (ICAR) Publication: New Delhi, 1985; pp 87–89.

13. Patel, N.; Rajput, T. B. S. Water and Nitrate Movement in Drip-irrigated Onion under Fertigation and Irrigation Treatments. *Agric. Water Manage.* **2006,** *79,* 293–311.

14. Pawar, D. D.; Dingre, S. K.; Shinde, M. G.; Kaore, V. K. *Drip Fertigation for Higher Crop Productivity*; MPKV Research Publication: Akola, India, 2013; pp 6–81.

15. Pawar, D. D.; Dingre, S. K.; Kale, K. D.; Surve, U. S. Economic Feasibility of Water Soluble Fertilizer in Drip Fertigated Tomato. *Indian J. Agri. Sci.*, **2013,** *83* (7), 703–707.

16. Rajaraman, G.; Pugalendhi, L. Potential Impact of Spacing and Fertilizer Levels on the Flowering, Productivity and Economic Viability of Hybrid Okra (*Abelmoschus esculentus* L. moench) under Drip Fertigation System. *Am. J. Plant Sci.* **2013,** *4,* 1784–1789.

17. Rani, R.; Nirala, S. K.; Suresh, R. Effect of Fertigation and Mulch on Yield of Pointed Gourd in North Bihar. *Environ. Ecol.* **2012,** *30* (3A), 641–645.

18. Sharma, S.; Patra, S. K.; Roy, G. B.; Bera, S. Influence of Drip Irrigation and Nitrogen Fertigation on Yield and Water Productivity of Guava. *The Bioscan* **2013,** *8* (3), 783–786.

19. Srinivas, K. Growth, Yield and Quality of Banana in Relation to Nitrogen fertilization. *Tropical Agric.* **1997,** *74,* 260–262.

20. Tiwari, K. N.; Mal, P. K.; Singh, R. M. Response of Okra (*Abelmoschus esculentus* (L.) Moench.) to Drip Irrigation under Mulch and Non-mulch Conditions. *Agric. Water Manage.* **1998,** *38* (2), 91–102.

CHAPTER 5

EFFECTS OF DIFFERENT NITROGEN LEVELS ON DRIP-IRRIGATED CUCUMBER UNDER GREENHOUSE CONDITIONS

S. K. PATTANAAIK and P. DEBNATH

ABSTRACT

Effects of different nitrogen levels were evaluated on drip-fertigated cucumber (*Cucumis sativus*) grown in greenhouse. The cucumber plants were grown in beds made inside the green house. The nitrogen was applied through drip irrigation at calculated volume to each plant in each treatment. Among the treatments, T_3 (17.8 mmol/L N_2) produced significant highest yield and highest number of fruits.

5.1 INTRODUCTION

Greenhouse cultivation has become common cultivation practice for commercial cultivation of vegetables under adverse climatic conditions. It has been proven that good-quality produce is obtained from greenhouse with application of inputs including nutrients and minerals. Green house cultivation involves high fertilizer inputs resulting in higher yields of vegetable crops in comparison with open field conditions.[8] However, the higher application of fertilizer causes leaching down of N and P and thereby, pollutes the groundwater.[9,10,11,13] North-East Region of India receives high rainfall at high intensity. Open cultivation of vegetable crops becomes a difficult task. Under such situations, greenhouse cultivation provides a feasible technology. One of the major advantages is control of microclimate and absence of torrential natural precipitation.[7]

For satisfactory yield of cucumber (*Cucumis sativus*), high temperature and adequate soil moisture are essential. Even if under favorable conditions, cucumber is subjected to common problems of reduction in number of female flowers,[1] delay in fruit growth,[3-5] and mineral disorder.[2] Therefore, planting is usually done during spring to summer season when the weather conditions are favorable for plant growth and high yield. The congenial environment of soil favors the roots to develop to their maximum potential. Cucumber is sensitive to waterlogged conditions. For its better growth, well-drained soil is required.

Nutrient management practices should be adopted to supply and maintain an optimum level of nutrient status within the crop root zone. This can be achieved by adopting drip/trickle irrigation. Application of water-soluble fertilizer in required dozes along with drip irrigation is known as fertigation, which provides high fertilizer use efficiency.[3] When properly managed, it opens up new avenues for growing plants under adverse conditions such as the peculiar climatic conditions of Pasighat, Arunachal Pradesh. Fertigation has been widely used in greenhouse cultivation of vegetable crops due to fertilizer saving to increase quality productivity.

Drip-irrigated crops has less root volume and thereby can conserve water and nutrient in the soil.[3] Since root-zone of crop is only wetted, nutrients and its application frequency are more important in fertigation than the conventional method of fertilization. Many experimental studies have been conducted to standardize the water and nutrient requirement for different crops. [1,3,4,6,7,9,10,14,19] Under agro-climatic conditions of Arunachal Pradesh, information on levels and frequency of nutrients via fertigation is scarce for vegetables grown in greenhouse environment.[18]

Therefore, this research study was conducted to standardize the nitrogen requirements of cucumber grown in greenhouse. This chapter discusses effects of nitrogen fertigation levels on performance of drip-irrigated cucumber under greenhouse conditions.

5.2   MATERIALS AND METHODS

5.2.1   STUDY AREA

The present investigation was carried out in the greenhouse located in the research farm of the College of Horticulture and Forestry, Central Agricultural University of Pasighat, East Siang district of Arunachal Pradesh. The geographical location is between 27.3–29.42°N latitude and 94.42–95.35°E longitude and at 155 m above mean sea level. Pasighat has a climate, which is tropical humid during summer and dry mild winter. The average annual rainfall of the East Siang district is 4510 mm distributed over the year. The major portion is received during May to September. However, the rain starts during the last week of March and continues through September. The period from October to March is dry with acute shortage of irrigation water.

The soil is highly porous, having low water holding capacity, high infiltration rate, and high bulk density, thus having adverse effects on soil–water–plant continuum. High permeability of soil causes low nutrient use because of downward migration of available plant nutrients away from the root zone. The soil is also highly acidic. The physiochemical properties of the soil are given in Table 5.1. Anionic plant nutrients like nitrogen (N), sulfur (S), and Boron (B) are more prone to leaching losses.

TABLE 5.1 Important Physiochemical Soil Properties.

Properties	Units	Value
pH		5.3
Organic carbon	g/kg	23.4
Sand	%	62
Silt	%	21
Clay	%	17
Bulk density	g/cm^3	1.45
Particle density	g/cm^3	2.58
Porosity	%	43.79
Water holding capacity	%	63.24
Texture	–	Sandy loam
CEC C	mol/kg	14.5
Available N$_2$	mg/kg	151
Bray's P	mg/kg	12.7
Available K	mg/kg	74
Available B	mg/kg	0.36
DTPA—extractable	mg/kg	0.78

5.2.2 *CULTIVATION OF CUCUMBER*

The soil in the beds was first sterilized by drenching with formalin (1%) at the rate of 0.5 L/m^2. Then farm yard manure (FYM) was applied at of 10 t/ ha. One month–old cucumber seedlings were planted during the first week of February of 2015 in the beds prepared inside the naturally ventilated greenhouse. Plants were pruned to a single stem by removing lateral shoots. All the recommended cultural practices were adopted from time to time. Harvesting was done starting at 55 days after planting. As a commercial practice, curved or deformed fruits were removed from the plant during pruning operations and marketable mature fruits were harvested in eight pickings at an interval of 2–3 days. The number of fruits produced per plant was recorded.

5.2.3   DRIP IRRIGATION LAYOUT

The cucumber was irrigated by drip irrigation system. There were eight beds and each bed had two rows of plants. The plant spacing in each row was 60 cm, and the row-to-row spacing was also maintained at 60 cm. Half of the area of each bed was irrigated by separate drip line as shown in Figure 5.1. Each drip line was provided with separate lateral valve. The laterals were provided with 2-L/h in-line drippers.

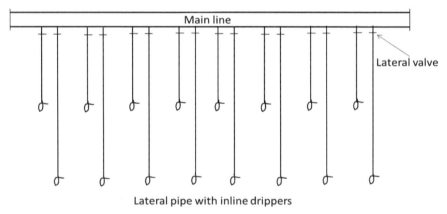

FIGURE 5.1   Layout of drip irrigation system inside the greenhouse.

The different levels of nitrogen were applied via fertigation using Venturi system. The stock solution was made in a 25-L PVC container. Water containing nitrogen as mentioned in Table 5.2 was applied to the respective plants under each treatment through a fertigation unit by controlling each lateral valve. Urea was used as the source of nitrogen. The irrigation schedule was maintained uniform in all the treatments, such as:

- <6 days of transplanting: 30 mL/plant/day,
- 7–25 days: 60 mL/plant/day,
- 26–55 days: 120 mL/plant/day,
- 56–90 days: 130 mL/plant/day,
- 91–110 days: 140 mL/plant/day.

The screen filter allowed removal of sediments in the irrigation water. Fertigation was before the solution enters the main and lateral pipe lines. During this period, a total amount of 376 L water was used per plant. The total cumulative amount of N applied was 0.75 kg, 1.50 kg, 2.26 kg, and 3.01 kg per plant in the treatments T_1, T_2, T_3, and T_4, respectively (Table 5.2).

TABLE 5.2   Different Treatments Used in the Experiment.

Treatment		Details
T_1	5.8 mmol/L N_2	4.35 g Urea in 25 L of water
T_2	11.8 mmol/L N_2	8.7 g Urea in 25 L of water
T_3	17.8 mmol/L N_2	13.05 g Urea in 25 L of water
T_4	23.8 mmol/L N_2	17.4 g Urea in 25 L of water

5.2.4   FERTIGATION

After realizing the importance of micro irrigation system, the next most essential input to boost the productivity and quality is the fertilizer. Fertigation is an application of fertilizer through an irrigation system. The fertilizers are applied daily or alternate day as per the requirement of crops to the root zone. The availability of 100% water-soluble solid (WSS) or liquid fertilizer is essential, if it is to be applied through fertigation. For fertigation, the special fertilizers are available in the following forms:

- *Liquid fertilizers* are the solutions that contain one or more plant nutrients in liquids form, but these are not popular in India.
- *WSS fertilizers* are 100% water-soluble fertilizers, and generally contain two or more major nutrients including micronutrients. These fertilizers completely dissolve in water leaving no precipitates. Therefore, there is no chance of clogging of emitters due to fertigation.

The most important criterion for the suitability of an irrigation system for fertigation is accuracy of water application, which largely depends on proper designing and installation of the irrigation system and availability of correct equipment for injecting fertilizers. The success of fertigation not only depends on good irrigation or proper use of equipments, but it also requires adequate information on crop water requirements, soil moisture characteristics, and crop physiology to enable accurate irrigation scheduling. Efficient fertigation requires access to information on soil fertility and nutrient balance in soil.

5.2.4.1   ADVANTAGES OF FERTIGATION

- The accurate placement of fertilizers in the active root zone with less loss through leaching.
- The application of nutrients in an available form can maintain the optimum level of nutrients required for crop growth. These two considerations offer potential for higher fertilizer application efficiency than with the conventional application method, resulting in 20–30% saving in fertilizer use.
- High-frequency fertigation avoids high peaks in salt concentration in the soil solution, thus maintaining a high osmotic potential and aiding nutrient uptake.
- Use of acidic fertilizers for fertigation on high-lime soils can reduce the pH within the zone, thus reducing the problems of nutrient fixation.
- Once desirable information is gathered regarding the peak requirement for nutrient uptake, fertigation can be scheduled to apply the correct proportion of nutrients for optimum yield. Timely application of fertilizer can help to optimize crop yield.
- The automation of fertigation can save in labor cost and reduces mechanical damage to crop and soil structure by laborers entering in the field.

5.2.4.2   FERTIGATION METHOD

In the beginning based on the desirable schedule, system has to run for 15–30 min then fertigation is done, again followed by irrigation for 15–30 min. During fertigation, it should be ascertained that the optimum concentration is maintained taking into consideration the crop reaction. For sensitive crops like tomato and strawberry, it should be about 500 ppm and for other crops maximum is 1000 ppm.

5.2.4.3   FERTILIZERS USED IN FERTIGATION

- *Straight fertilizers*: Granular urea, urea phosphate (17:44:0), ammonium phosphate (12:61:0), potassium phosphate (0:52:34), potassium nitrate (13:0:46), sulfate of potash (0:0:50 + 18%S), calcium nitrate, magnesium nitrate, and calcium magnesium nitrate.
- *NPK formulations*: 13:40:13 + TE, 20:20:20, 6:12:36, 13:05:26, 19:19:19 + TE, 16:08:24 + TE + Mg, and 20:10:10 + Trace elements (TE) + Mg.
- *Chelated micronutrients*: Fe-EDTA (13%), Zn-EDTA (15%), Fe-EDDH-A (6%), chelated combination of micronutrients, etc.

5.2.5   OBSERVATIONS RECORDED

The yield was recorded starting on March 23, 2015 through May 12, 2015. At the time of each harvesting, the yield from each plant was noted and the yield of fruits per treatment was calculated. The yield was recorded in terms of numbers of fruits and their weight in kg (Table 5.3).

TABLE 5.3   Yield per Plant of Drip-Irrigated Cucumber under Greenhouse Conditions.

Treatments	Fruit yield (kg)	Number of fruits
T_1	16.850	131
T_2	18.631	141
T_3	24.181	174
T_4	15.388	128
CV	**30.370**	**24**
CD at p = 5%	**5.926**	**36**

5.2.6   STATISTICAL ANALYSIS

The data was analyzed by WASP 1.0 statistical software developed by ICAR, Goa.

5.3   RESULTS AND DISCUSSION

The study revealed that the threshold N level for greenhouse cultivation of cucumber under Pasighat climatic condition was 17.8 mmol/L of N_2. Beyond this level of N in soil, there was decrease in yield and fruits per plant. The maximum fruit yield was recorded 24.18 kg/plant for the treatment T_3 (17.8 mmol/L N_2). The highest number of fruits per plant (174.25) is also recorded for T_3. The yield and fruits per plant in T_3 were significantly higher than the treatments T_1, T_2, T_4 (Table 5.2). However, the decrease in yield and fruits per plant in the treatment T_4 may be attributed to the toxic effects of nitrogen. Figures 5.2–5.4 indicate of cucumber growth under greenhouse conditions. The Figures 5.5 and 5.6 indicate that the fruit yield has positive correlation with number of fruits with a correlation coefficient of 0.98. This is in accordance to the finding by other investigators.[6]

FIGURE 5.2   Trailing of cucumber plant along with drip fertigation line.

5.4   SUMMARY

This chapter discusses response of fertigation on performance of cucumber. The treatments were T_1—5.8 mmol/L N_2, T_2—11.8 mmol/L N_2, T_3—17.8 mmol/L N_2, and T_4—23.8 mmol/L N_2. The nutrient was applied through drip irrigation at calculated volume to each plant in each treatment. FYM at 10 t/ha was applied in the soil at the time of bed preparation. During this period, a total amount of 376 L of water was applied per plant. Application of nitrogen through drip irrigation showed a positive effect on the fruit yield and number of fruits per plant. The treatment T_3 (17.8 mmol/L N_2) produced significant highest yield and highest number of fruits compared with other treatments. The yield was decreased with higher level of N_2 application in treatment T_4 (23.8 mmol/L N_2).

FIGURE 5.3   Fruit production of cucumber.

FIGURE 5.4   Fertigation trial in greenhouse cultivated cucumber.

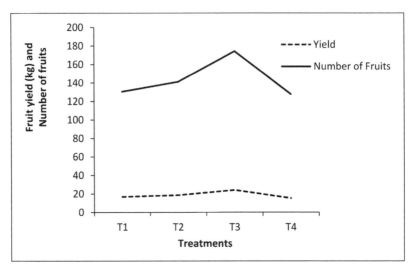

FIGURE 5.5   Effects of nitrogen treatments on fruit yield and number of fruits.

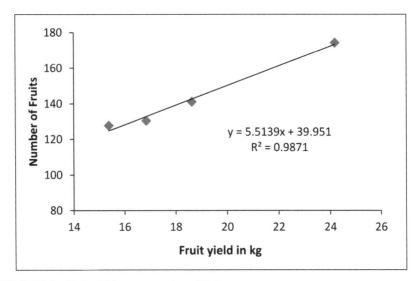

FIGURE 5.6   Fruit yield versus number of fruits.

ACKNOWLEDGMENT

The authors are thankful to: the Central Agricultural University (CAU)—Imphal, Manipur, to provide facilities to conduct the experiment; ICAR—Goa for online software.

KEYWORDS

- cucumber
- drip irrigation
- greenhouse
- nitrogen fertigation

REFERENCES

1. Albregt, E. E.; Hochmutch, G. J.; Chandler, C. K. Potassium Fertigation Requirements of Drip Irrigated Strawberry, *J. Am. Soc. Hortic. Sci.* **1996,** *121,* 164–168.
2. Bakker, J. C.; Sonneveld, C. Calcium Deficiency of Glasshouse Cucumber as Affected by Environmental Humidity and Mineral Nutrition. *J. Hortic. Sci.* **1988,** *63* (2), 241–246.
3. Bar-Yosef, B.; Stammers, C. Growth and Trickle Irrigated Tomato as Related to Rooting and Uptake of N and Water. *Agron. J.* **1980,** *72,* 815–822.
4. Bhella, H. J.; Wilcox, G. E. In *Nitrogen Fertilization and Muskmelon Growth, Yield and Nutrition,* Drip/Trickle irrigation Action Proceeding for Third International Drip/Trickle Irrigation Congress, November 18–21, 1985; ASABE: St Joseph, MI, California, USA, 1985; pp 339–344.
5. Cantliffe, D. J. Alteration of Sex Expression in Cucumber Due to Changes in Temperature, Light Intensity, and Photoperiod. *J. Am. Soc. Hortic. Sci.* **1981,** *106* (2), 133–136.
6. Hartz, T. K. Drip Irrigation Scheduling for Fresh Market Tomato Production. *Hortic. Sci.* **1993,** *23,* 35–37.
7. Horton, R.; Beese, F.; Wierenga, P. J. Physiological Response of Chilli Pepper to Trickle Irrigation. *Agron J.* **1982,** *74,* 551–555.
8. Liebig, H. P. Physiological and Economical Aspects of Cucumber Crop Density. *Acta Hortic.* **1981,** *118,* 149–164.
9. Locasio, S. J.; Hochmuch, G. J. Nitrogen and Potassium Application Scheduling Effects on Drip Irrigated Tomato Yield and Leaf Tissue Analysis. *Hortic. Sci.* **1997,** *32,* 230–235.
10. Locasio, S. J.; Oslon, S. M. Water Quality and Time of N & K Application for Trickle Irrigated Tomatoes. *J. Am. Soc. Hortic. Sci.* **1989,** *114,* 265–268.
11. Marcelis, L. F. M.; Hofman-Eijer, L. R. Effect of Temperature on the Growth of Individual Cucumber Fruits. *Physiol. Plant.* **1993,** *87* (3), 321–328.
12. Medany, M. A.; Wadid, M. M.; Abou-Hadid, A. F. Cucumber Fruit Growth Rate in Relation to Climate. *Acta Hortic.* **1999,** *486,* 107–111.
13. Papadorpoulos, I. Nitrogen Fertigation of Greenhouse-grown Cucumber. *Plant Soil* **1986,** *93* (1), 87–93.
14. Papadorpoulos, I. Nitrogen Fertigation of Greenhouse-grown Straberries. *Fert. Res.,* **1987,** *13,* 269–276.
15. Ruiji, M. N. A. Economic Evaluation of Closed Production Systems in Greenhouse Horticulture. *Acta Hortic.* **1995,** *340,* 87–94.

16. Sonneveld, C. *Mineral enbalansen bij kasteelten, Meststoffen* (Mineral enbalances at Manor, Fertilizers). *NMI Wageningen* **1993,** *93,* 44–49.
17. Sonneveld, C. Effects of Salinity on Substrate Grown Vegetables and Ornaments in Green House Horticulture. Ph.D. Dissertation, Wageningen University, Wageningen, 2000; p 151.
18. Thomas, L. T.; White, S. A. Fertigation Frequency for Sub-surface Drip Irrigated Broccoli. *Soil Sci. Soc. Am. J.* **2013,** *67,* 910–918.
19. Wierenga, P. J.; Hendrickx, J. M. Yield and Quality of Trickle Irrigated Chilli Peppers. *Agric. Water Manage.* **1985,** *9,* 339–356.
20. Wunderink, H. *De bevelasting van het Nederlandse oppervlaktewater met fosfat en stikstof* (The Burden on the Dutch Surface Water with Phosphate and Nitrogen). *Het Waterschap* (Water Board, Netherlands) **2017,** *81* (9), 304–313.

CHAPTER 6

PERFORMANCE OF SUNFLOWER WITH DIFFERENT IRRIGATION METHODS: COASTAL PLAIN ZONE OF EASTERN INDIA

ARATI SETHI, NARAYAN SAHOO, BALRAM PANIGRAHI,
BENUKANTHA DASH, and LALA I. P. RAY

ABSTRACT

Irrigation is one of the major key factors to address the needs of food, fiber, and shelter. To maximize grain production, the irrigation water-use efficiency should be enhanced. Sunflower is a water-sensitive crop. This chapter evaluated its performance under drip, sprinkler, and furrow irrigation for soil moisture deficit conditions. The water requirement under drip system was 415.8 mm, with an average yield of 1.31 t/ha, compared to 1089.8 mm and yield of 1.1 t/ha under furrow system.

6.1 INTRODUCTION

More than 65% of Indian rural population depends on agriculture for the livelihood security. Most of the increase in food grain production will be met by increasing the irrigated area and irrigation efficiency in response to water scarcity. In India, >70% of available water resources are used for irrigation purposes.

Within the limited water resources and increasing demand for water, identification of suitable irrigation methods and optimum supply of water to avoid water stress during growing period is necessary for maximum yield. Therefore, a field experiment on hybrid sunflower has been conducted by various researchers to identify the efficacy of different irrigation methods.

Application of timely and appropriate quantity of irrigation water during critical growth stages increases the productivity of sunflower crop considerably.[2]

Subsurface drip irrigation system has resulted in higher seed yield and water-use efficiency (WUE) by applying irrigation water at 125% of ET_0.[6] Amount of irrigation water computed through modified Blaney–Criddle equation is much closer to the actual amount of water applied in the sunflower crop.[7] Yield of drip-irrigated sunflower is 26% more, required 56% less water and WUE was three times higher than the furrow irrigation.[5,8]

Grain yield is significantly reduced due to delay in sowing date of sunflower crop.[1] Yield reduces due to water stress during seed formation compared to 100% irrigation, and less reduction was observed for stress during flowering stage.[10,11] Both seed yield and oil yield were greatly influenced by quantity and distribution of water.[9] Oil yield of sun flower was significantly increased by providing optimum quantity water during flowering stage.[3,4] Crop yield is higher, when sufficient irrigation water was provided in flowering and seed formation stages.[10]

Due to water scarcity and irrigation related issues, the situation demands to adopt the best irrigation method compared to the traditional gravity irrigation to obtain optimum crop yield. Combination of different irrigation methodologies with different levels of irrigation for various crops is required to identify the best irrigation method, which can give optimum yield. Hence, a field experiment was conducted for sunflower crop as a test crop.

The objectives of this research on sunflower are to evaluate the effects of drip, micro-sprinkler, and furrow irrigation methods with different levels of irrigation on crop yield and other biometric parameters.

6.2 METHODOLOGY

6.2.1 IRRIGATION METHODS

In this research, sunflower crop under different irrigation methods and different amounts of irrigation at different maximum allowable deficit (MAD) levels was used to identify the best management practices with maximum yield. A view of experimental plots from field preparation to crop maturity is shown in Figures 6.1–6.4. Details of different irrigation methods with different irrigation amounts at different MAD levels are presented in Table 6.1.

FIGURE 6.1 Layout of the drip system.

FIGURE 6.2 Field observations in progress.

FIGURE 6.3   Sunflower plot after irrigation.

FIGURE 6.4   Sunflower crop at maturity.

TABLE 6.1 List of Treatments: Irrigation Methods at Different Maximum Allowable Deficit (MAD) Levels.

Irrigation levels	Irrigation methods		
	Drip (I_1)	Sprinkler (I_2)	Furrow (I_3)
	(% MAD)		
T_1	20	20	20
T_2	30	30	30
T_3	40	40	40
T_4	50	50	50

6.2.2 ESTIMATION OF IRRIGATION AMOUNT

6.2.2.1 ESTIMATION OF CROP EVAPOTRANSPIRATION

The FAO CROPWAT-8.0 was used for calculation of daily crop evapotranspiration of sunflower crop using parameters, such as climate, crop, and soil data.

6.2.2.2 ESTIMATION OF IRRIGATION WATER REQUIREMENT

Before conducting the experiment, field capacity and bulk density of soil at the experiment site were determined. Daily moisture was measured from each plot using digital soil moisture meter. Then amount moisture utilized as ET_{crop} was computed by differentiating soil moisture at field capacity and moisture measured at different MAD levels through digital moisture meter. Then, the decreased amount of soil moisture was replenished by irrigation water using drip, micro-sprinkler, and furrow irrigation, when the soil moisture was decreased by 20, 30, 40, and 50% of the field capacity.

6.2.2.3 BIOMETRIC PARAMETERS

Different irrigation methods with a combination of different amounts of irrigation can directly or indirectly affect the biometric parameters. Plant height at different stages of growth, root length (at both vertical and lateral distances), stem diameter, diameter of head, leaf area index (LAI), test weight of 1000 grains, and the yield was computed for each irrigation method. SPSS 16.0 software was used for analysis of yield and yield attributes for various irrigation methods and levels of irrigation.

6.2.2.4   WATER-USE EFFICIENCY

The ratio between crop yield (kg/ha) to the total amount of water delivered to field is defined as WUE, which can be calculated using the following formula:

$$WUE = \frac{Crop\ yield\ in\ kg\ per\ hectare}{Total\ water\ delivered,\ mm} \tag{6.1}$$

6.3   RESULTS AND DISCUSSION

6.3.1   COMPUTATION OF WATER REQUIREMENT OF SUNFLOWER

6.3.1.1   IRRIGATION WATER REQUIREMENT

During the crop period, the irrigation water required (IWR) for various treatments is presented in Table 6.2. The maximum irrigation requirement at 20% MAD level (i.e., 1044 mm) was observed in furrow irrigation compared to other MAD levels.

TABLE 6.2   Irrigation Water Requirement (mm) of Sunflower under Different Treatments.

Irrigation levels	Irrigation water requirement (mm)		
	I_1	I_2	I_3
T_1	503.0	563.0	1044.0
T_2	468.0	528.0	934.0
T_3	419.0	482.0	774.0
T_4	370.0	430.0	614.0

6.3.1.2   WATER REQUIREMENT

The total amount of IWR for sunflower during the growing period was estimated by considering effective rainfall (ER), water applied, and contribution of groundwater. Out of seven rainfall events, the ER amount was 45.8 mm during the crop growth. The deep percolation (DP), soil moisture contribution (SMC), etc. were calculated for each treatment. Water requirement (WR) of sunflower is presented in Table 6.3.

TABLE 6.3 Water Requirement (mm) of Sunflower under Various Irrigation Levels and Irrigation Methods.

Treatment	Irrigation level	IWR	ER	SMC	DP	WR
I_1	T_1	503	45.8	0	0	548.8
	T_2	468	45.8	0	0	513.8
	T_3	419	45.8	0	0	464.8
	T_4	370	45.8	0	0	415.8
I_2	T_1	563	45.8	0	0	608.8
	T_2	528	45.8	0	0	573.8
	T_3	482	45.8	0	0	527.8
	T_4	430	45.8	0	0	475.8
I_3	T_1	1044	45.8	0	0	1089.8
	T_2	934	45.8	0	0	979.8
	T_3	774	45.8	0	0	819.8
	T_4	614	45.8	0	0	659.8

It is observed that the total WR of the sunflower crop is 1089.8 mm at 20% MAD level with furrow irrigation, which is the highest compared to other irrigation methods and other irrigation levels. The minimum total WR of 415.8 mm was under drip irrigation method with 50% MAD.

6.3.2 ESTIMATION OF EVAPOTRANSPIRATION OF SUNFLOWER

Daily evapotranspiration (ET_0) rate was computed by using soil, climate, and crop data in CROPWAT-8.0, and the daily values for different crop stages and irrigation WR for each stage are shown in Table 6.4. There was an increase in evapotranspiration rate when mean wind velocity and temperature were increased. During February and March, evapotranspiration rate was low and was gradually increased in April and May. This may be due to lower wind velocity and temperature during February and March. Differences in actual and estimated irrigation WR were higher in furrow irrigation than those under drip and sprinkler irrigation methods. This may due to increasing soil porosity and more evaporation loss due to weeding, intercultural operation, and water consumed by weeds before weeding operation.

TABLE 6.4   Crop Stage Wise ET_c and Irrigation Requirement.

Month	Date	Crop stage	Crop coefficient (K_c)	ET_c	ET_c	Effective rain	Irrigation requirement
			–	mm/day		mm/stage	
February	3	Initial	0.35	1.38	9.6	2.6	7.0
March	1	Development	0.36	1.6	16.0	0.0	16.0
March	2	Development	0.65	3.25	32.5	1.3	31.2
March	3	Mid	1.05	5.75	63.2	0.2	63.0
April	1	Mid	1.15	7.01	70.1	0.0	70.1
April	2	Mid	1.15	8.15	81.5	0.0	81.5
April	3	Late	1.13	7.96	79.6	7.3	72.3
May	1	Late	0.83	5.64	56.4	2.5	53.9
May	2	Late	0.48	3.37	27.0	25.7	1.3
				Total	435.9	39.6	396.3

6.3.3   GROWTH, YIELD, AND YIELD ATTRIBUTES

6.3.3.1   GROWTH PARAMETERS

6.3.3.1.1   Plant Height

Mean height of plant for different irrigation methods for various crop stages with standard deviation is presented in Table 6.5 and Figure 6.5.

At 20% MAD level in furrow irrigation methods, the highest mean height was 17.5 cm for the development stage, 104.35 cm for mid-season stage, and 115 cm for late-season stage. The highest plant height in furrow irrigation method may be due to requirement of more water and intercultural operation (loosening of soil and weeding). Under micro-sprinkler treatments, the plant height was observed to be comparatively lower due to nutrient leaching from the root zone. Statistical analysis showed that the differences among levels and methods of irrigation for initial-, mid-, and late-season stages were significant.

6.3.3.1.2 Root Length

The relationship between irrigation levels of three irrigation methods and growth of vertical and lateral roots is shown in Figure 6.6. The maximum mean vertical length of root was found in furrow irrigation, whereas the highest mean length of lateral root was observed in drip irrigation methods (Table 6.6 and Figure 6.6). The mean vertical length of root in furrow irrigation (I_3) was higher than the other irrigation methods at corresponding MAD irrigation levels. Due to the controlled application of irrigation water in drip and sprinkler irrigation methods, water is available at shallow depth, which increases the lateral growth of root, whereas vertical growth of the root was controlled due to less vertical movement of water. Therefore under drip irrigation, length of lateral root was higher. Statistical analysis showed that differences among levels and methods of irrigation relating to mean length of root were significant.

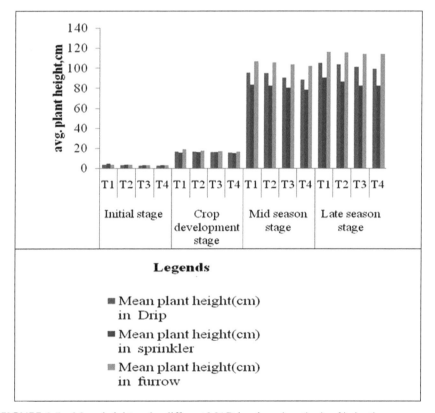

FIGURE 6.5 Mean height under different MAD levels and methods of irrigation.

TABLE 6.5 Mean Height (cm) for Different Crop Stages under Various Irrigation Methods.

Crop stage	I_1		I_2		I_3	
	Avg.	Sd	Avg.	Sd	Avg.	Sd
Initial	2.85	0.42	3.45	0.81	3.27	0.67
Crop development	16.05	0.96	15.52	0.94	17.50	2.31
Mid-season	91.87	5.76	80.75	5.26	104.35	13.50
Late-season	102.12	7.23	85.07	7.13	115.00	6.77

I_1: drip irrigation method; I_2: micro-sprinkler method; I_3: furrow irrigation method; Avg.: average; Sd: standard deviation.

TABLE 6.6 Mean Length of Root under Different Methods of Irrigation.

Irrigation method	Length of vertical root (cm)		Lateral root length (cm)	
	Avg.	Sd	Avg.	Sd
Drip	11.55	1.40	15.19	1.78
Sprinkler	11.01	1.17	11.92	1.25
Furrow	13.25	2.05	11.52	1.10

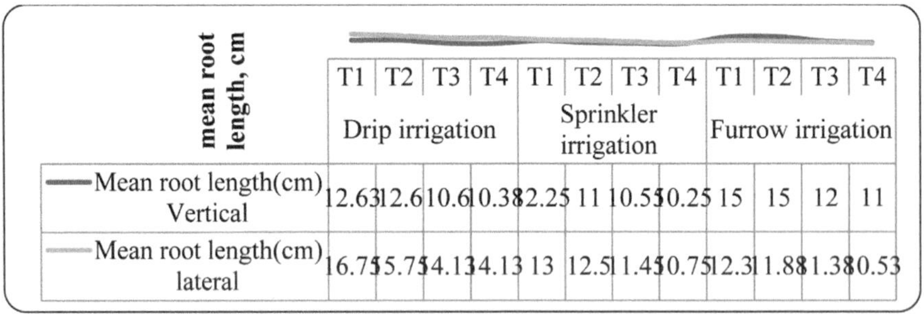

mean root length, cm	T1	T2	T3	T4	T1	T2	T3	T4	T1	T2	T3	T4
	Drip irrigation				Sprinkler irrigation				Furrow irrigation			
—Mean root length(cm) Vertical	12.63	12.6	10.6	0.38	2.25	11	10.55	0.25	15	15	12	11
—Mean root length(cm) lateral	16.75	5.75	4.13	4.13	13	12.5	1.45	0.75	12.3	1.88	1.38	0.53

FIGURE 6.6 Mean length of roots under different MAD levels and methods of irrigation.

6.3.3.1.3 *Stem Diameter*

The mean stem diameter in all the irrigation methods ranged between 6.92 and 6.77 cm showing negligible differences (Table 6.7 and Fig. 6.7). Statistical analysis showed that differences among irrigation levels and methods of irrigation relating to mean diameter of stem were nonsignificant.

TABLE 6.7   Mean Girth Diameter under Various Methods of Irrigation.

Irrigation methods	Avg. diameter of stem (cm)	Standard deviation
Drip	6.77	0.927
Sprinkler	6.55	1.190
Furrow	6.92	0.811

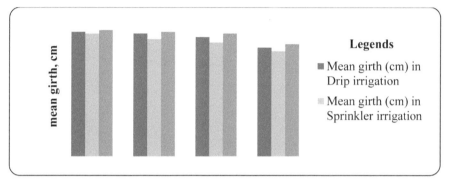

FIGURE 6.7   Mean diameter of stem under different levels and methods of irrigation.

6.3.3.1.4   Head Diameter

The highest mean diameter of head (16.60 cm) of sunflower was under drip irrigation method and the lowest was 13.5 cm under sprinkler irrigation methods (Table 6.8). The average maximum sunflower head diameter drip irrigation method was 16.60, 16.10, 15.90, and 15.50 cm, respectively at 20, 30, 40, and 50% MAD levels (Fig. 6.8).

TABLE 6.8   Mean Diameter of Head under Various Methods of Irrigation.

Methods of irrigation	Mean diameter of head (cm)	Sd
I_1	16.02	1.41
I_2	13.00	1.39
I_3	13.75	1.62

6.3.3.1.5   Leaf Area Index

Mean maximum LAI (3.34) was under furrow irrigation method compared to sprinkler and drip irrigation method (Table 6.9 and Fig. 6.9).

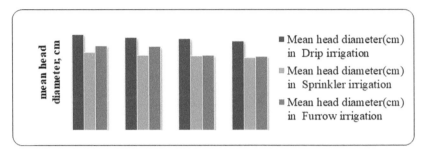

FIGURE 6.8   Mean head diameter under various levels and methods of irrigation.

TABLE 6.9   Mean Leaf Area Index in Various Methods of Irrigation.

Methods of irrigation	Mean LAI	Standard deviation
I_1	2.50	0.73
I_2	1.75	0.64
I_3	3.34	0.57

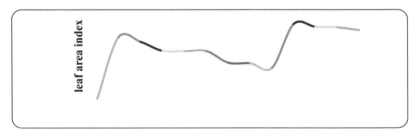

FIGURE 6.9   Mean leaf area index (LAI) under various levels and methods of irrigation.

6.3.3.1.6   *Test Weight*

The observed value of weight of 1000 grains of sunflower under various treatments is shown in Figure 6.10 and in Table 6.10. The test weight was higher under drip irrigation compared to the other methods of irrigation, due to better availability of water and nutrient. Statistical analysis showed that differences among irrigation methods and levels of irrigation were significant.

6.3.3.1.7   *Yield*

Sunflower yield data are presented in Table 6.11 and Figure 6.11. Yield variations were observed both under levels of irrigation and methods of irrigation. Among various treatments, the highest yield was 1301 kg/ha for drip-irrigated crop at 20% MAD, which was comparatively higher than the other irrigation methods, whereas the lowest yield was for furrow-irrigated crop at 50% MAD (1100 kg/ha).

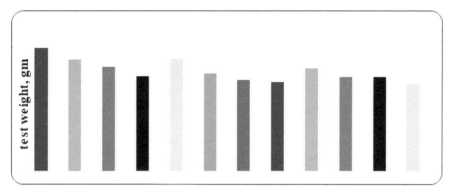

FIGURE 6.10 Test weight (g/1000 grains) under different levels and methods of irrigation.

TABLE 6.10 Mean Test Weight (g) under Various Methods of Irrigation.

Methods of irrigation	Mean test weight (g)	Standard deviation
I_1	44.330	7.638
I_2	39.965	4.222
I_3	38.725	2.790

TABLE 6.11 Crop Yield (kg/ha) under Various Methods of Irrigation.

Methods of irrigation	Mean yield (kg/ha)	Sd
I_1	1221.3	1.110
I_2	1196.3	1.025
I_3	1175.0	1.010

Multiple regression analysis showed that crop productivity is a function of stem girth, plant height at mid-season stage, diameter of head, lateral root length, and LAI. The growth parameters did not significantly contribute to the crop yield, and these parameters (height of plant during initial, crop development, and late-season stage; vertical length of root and test weight) are not significantly correlated to the crop productivity. The following linear regression equation was developed between yield (Y) and five morphological parameters of sunflower crop.

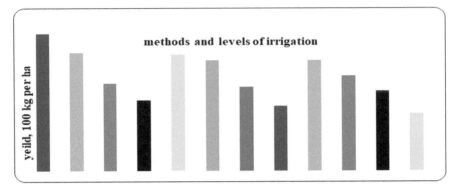

FIGURE 6.11   Yield (100 kg/ha) under various levels and methods of irrigation.

$$Y = 10.578 - (0.004 \times a) - (0.008 \times b)$$
$$- (0.078 \times c) - (0.111 \times d) + (0.247 \times e) \qquad (6.2)$$

where Y is the yield of sunflower crop (100 kg/ha); a is the plant height at mid-season stage, cm; b is the stem diameter, cm; c is the head diameter, cm; d is the LAI; and e is the lateral root length, cm.

6.3.3.1.8 Water-Use Efficiency

Maximum WUE was 2.528 kg/ha mm at 50% MAD level under drip irrigation method (Table 6.12 and Figure 6.12), due to influence of water management practices.

TABLE 6.12   Water-Use Efficiency under Various Methods of Irrigation.

Methods of Irrigation	Avg. WUE (kg/ha mm)	Sd
Drip	2.528	0.541
Sprinkler	2.198	0.575
Furrow	1.363	0.427

6.3.4 BENEFIT–COST RATIO FOR DIFFERENT IRRIGATION METHODS

Under drip irrigation method, crop yield was significantly higher compared to sprinkler and furrow irrigation methods. Detailed cost of cultivation for three methods of irrigation for all treatments is shown in Table 6.13.

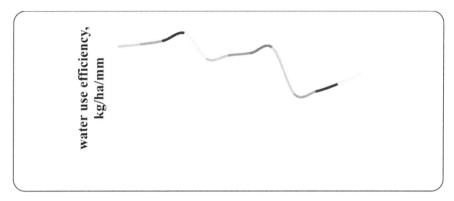

FIGURE 6.12 Water-use efficiency (WUE, kg/ha mm) under various levels and methods of irrigation.

TABLE 6.13 Analysis of Benefit–Cost Ratio under Different Irrigation Methods.

S.No.	Input parameter	I_1	I_2	I_3
1.	Cost of planting and seed, including land preparation, Rs./ha	25,000.00	250,00.00	33,333.00
2.	Cost of micro irrigation system, including installation (life span—10 years, nos. of crops—3/year), Rs./ha	16,666.00	13,125.00	–
3.	Cost of farm yard manure, Rs./ha	3708.00	3708.00	3708.00
4.	Cost of chemicals and pesticides, Rs./ha	931.00	931.00	931.00
5.	Cost of manual weeding, Rs./ha	3472.00	3472.00	5417.00
6.	Cost fertilizer, Rs./ha (manual application)	1389.00	1389.00	1389.00
7.	Cost of pumping charge for irrigation, Rs./ha	2778.00	3889.00	4167.00
8.	Cost of harvesting, Rs./ha	2750.00	2750.00	2750.00
9.	Cost of cultivation (sum of items 1–9), Rs./ha	56,694.00	54,264.00	51,695.00
10.	Water saving cost in furrow irrigation, Rs./ha	54,490.00	47,950.00	–
11.	Yield, kg/ha	1310.00	1255	1240
12.	Selling price of sunflower seed, Rs./kg	45.00	45.00	45.00
13.	Realization (gross benefit), Rs./ha	58,950.00	56,475.00	55,800.00
14.	Total realization cost, Rs./ha	113,440.00	104,425.00	55,800.00
15.	Net profit, Rs./ha	56,746.00	52,161.00	4105.00
16.	B:C ratio	2.00	1.92	1.08

Under drip irrigation system, net return (Rs./ha) is higher than the other two methods of irrigation. The benefit–cost ratio under drip irrigation was higher than other methods of irrigation. The net profit was highest (Rs. 56,746.00/ha) under drip irrigation followed by sprinkler irrigation method, which was 92.8% higher than the furrow irrigation. The realization cost per ha under drip irrigation was maximum (Rs. 113,440.00/ha). The benefit–cost ratio was highest (2.00) compared with furrow and sprinkler irrigation methods.

6.4   SUMMARY

During the cropping season, computed crop evapotranspiration was approximately 436.0 mm and irrigation WR is 396.0 mm. The maximum irrigation WR is 1044 mm under furrow irrigation at 20% MAD and the minimum irrigation requirement under drip method at 50% MAD is 370 mm; whereas the total WR is minimum for drip method at 50% MAD (i.e., 415.8 mm) and maximum was for controlled furrow method at 50% MAD (i.e., 1089.8 mm). The average yield of 1310 kg/ha was maximum under drip irrigation at 20% MAD level and minimum for furrow irrigation (1100 kg/ha). The WUE was maximum under drip irrigation at 50% MAD (2.73 kg/ha mm).

ACKNOWLEDGMENT

The corresponding author thanks OUAT authorities for all logistics during the tenure of the experiment; B. Dash and Lala I. P. Ray for data analysis.

KEYWORDS

- **benefit–cost**
- **crop stage**
- **evapotranspiration**
- **irrigation method**
- **irrigation requirement**

REFERENCES

1. Flagella, Z.; Rotundo, T., Tarantino, R. Change in Seed Yield and Oil Fatty Acid Composition of High Oleic Sunflower Hybrids in Relation to the Sowing Date and Regimes. *Res. J. Biol. Sci.* **2002,** *17*, 221–230.
2. Ghani, A.; Hussain, M.; Qureshi, M. S. Effect of Different Irrigation Regimes on the Growth and Yield of Sunflower, *Int. J. Agric. Biol.* **2000,** *2*, 334–335.
3. Goksoy, A. T.; Demir, A. O.; Turan, Z. M.; Dagustu, N. Response of Sunflower (*Helianthus annuus* L.) to Full and Limited Irrigation at Different Growth Stages. *Field Crops Res.* **2004,** *87* (2–3), 167–178.
4. Gyori, Z.; Nemeskeri, M. Legume Grown under No Irrigated Condition. *J. Agric. Food Chem.* **1998,** *46*, 3087–3091.
5. Hanson, B.; May, D. The Effect of Drip Line Placement on Yield and Quality of Drip Irrigated Processing Tomatoes. *Irrig. Drain. Syst.* **2007,** *21*, 109–118.
6. Mehana, H. I. Maximizing Water Use of Sunflower Crop under Arid Eco Systems. M.Sc. Thesis; Faculty of Agriculture, Ain Shams University, Cairo, Egypt, 2005; p 152.
7. Nahla, A. H. Water Regime of Some Crops Grown under Drip Irrigation at El-Kharga Oasis. M.Sc. Thesis; Agric. Sci. Soils Department, Assiut University, Egypt, 2003; p 115.
8. Qureshi, A. L.; Gadehi, M. A.; Mahessar, A. A. Effect of Drip and Furrow Irrigation Systems on Sunflower Yield and Water Use Efficiency in Dry Area of Pakistan. *J. Agric. Environ. Sci.* **2015,** *15* (10), 1947–1952.
9. Reddy, G. K. M.; Dangi, K. S.; Kumar, S. S.; Reddy, A. V. Effect of Moisture Stress on Yield and Quality in Sunflower (*Helianthus annuus* L.). *J. Oilseeds Res.* **2003,** *20* (2), 282–283.
10. Stone, L. R.; Schlegel, A. J.; Gwi, R. E.; Khan, A. H. Response of Corn, Grain, Sorghum and Sunflower to Irrigation in the High Plains of Kansas. *Agric. Water Manage.* **1996,** *30*, 251–259.
11. Tolga, E.; Lokman, D. Yield Response of Sunflower to Water Stress under Tekirdag Conditions. *Helia* **2003,** *26* (38), 149–158.

SUMMER *KUSMI* LAC PRODUCTION ON DRIP IRRIGATED *FLEMINGIA SEMIALATA* ROXB

R. K. SINGH

ABSTRACT

Lac is one of the forest products that are mostly harvested under wild conditions. It is the source of livelihood for the tribal people inhabited in forest and nearby regions. However, the same lac can be produced commercially under certain cultivation practices. *Flemingia semialata* Roxb. (Papilionaceae), host plant, has shown promise for significant production of lac due to its fast and vigorous growth with high coppicing response on plantation basis. *F. semialata* has been very suitable for good-quality production of *kusmi* lac crop (*aghani*) during the winter season. However, performance of crop during the summer season (*jethwi*) is very poor due to high temperature. Hence, the performance of lac production was studied under irrigated conditions during summer season. Significant yield was achieved under irrigated conditions compared with nonirrigated plants. Broodlac harvested for irrigated plants was 150 g/plant compared to 10 g/plant for nonirrigated plants. Therefore, broodlac yield ratio was 5:1 and 0.7:1 under these conditions, respectively.

7.1 INTRODUCTION

Lac is a natural nontoxic resinous produce of an insect belonging to genus *Kerria* (Homoptera: Tachardiidae). The Indian lac insect *Kerria lacca* (Kerr) is cultured mainly on *palas* (*Butea monosperma*), *kusum (Schleichera oleosa)*, and *ber (Ziziphus mauritiana)*, having two strains, *kusmi* and *rangeeni*, and both are bivoltine in nature. The former is known for its superior quality and productivity of lac resin. Long gestation period (time period to be able to take lac insect inoculation) of these conventional lac host trees is the bottleneck for the lac growers, who do not possess these tree species or for those, who want to take up intensive lac cultivation for assured income.

Out of many quick growing bushy lac hosts identified for intensive lac cultivation, *Flemingia semialata* Roxb. (Papilionaceae) has shown promise due to its fast and vigorous growth with high coppicing response on plantation basis.[2,3,5] This host is suitable for good-quality winter season *kusmi* lac (*aghani*) crop production. However, summer season (*jethwi*) crop does not thrive well due to high temperature, which coincides with the crop maturing stage.

Keeping this in view, a study was undertaken on summer *kusmi* lac (*jethwi*) crop production under drip irrigation system. The hypothesis behind this study was that irrigation to the bushy lac host plant will keep the soil moist around the plant periphery, and desiccating shoots of the plants will never be short of moisture and nutrients supply that are necessary for plant metabolic activities in the successful lac production. Enhanced summer season sticklac yield under pitcher irrigation over control (no irrigation) for 2 consecutive years was reported in a study conducted at IINRG (Indian Institute of Natural Resins and Gums), Ranchi.[4]

This chapter focuses on summer *kusmi* lac production on drip-irrigated *F. semialata* Roxb.

7.2   MATERIALS AND METHODS

The study was undertaken at IINRG—Namkum Ranchi Research Farm (23°23′N, 85°23′E and 652 m msl) from July 2006 to August 2007. Sixty *semialata* plants were selected for this study, out of which 30 plants were kept under control (nonirrigated) and 30 under irrigated conditions. Plants were inoculated by tying bundles of 15 g of broodlac (lac encrustation carrying gravid females) per bush in the third week of February for raising summer season *kusmi* lac crop. The plant-to-plant and row-to-row spacing was 1 m × 1 m.

The irrigation was provided through drip method on alternate day for 1 h during 15th March to 15th June. There was one dripper for every plant and the discharge of dripper was 4 L/h.

All the management practices and insect-pest management were carried out by using the recommended practices for *kusmi* crop.[1] The parameters for study were plant height (cm) and diameter (cm) before lac inoculation and at the time of harvesting the lac crop. The crop was harvested in late August and the data were recorded for broodlac production and broodlac yield ratio.

7.3   RESULTS AND DISCUSSION

7.3.1   EFFECT ON PLANT GROWTH ATTRIBUTES

Plant height and diameter showed a definite increase under irrigated conditions. Plant height under irrigated and nonirrigated conditions was 134.3 and 104 cm, while plant diameter was 67.7 and 39.3 cm, respectively (Fig. 7.1). Thus, ratio for an increase in plant height and diameter was 1.3 and 1.7 times under irrigated and nonirrigated conditions, respectively.

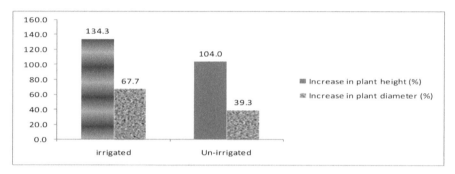

FIGURE 7.1   Growth parameters of *F. semialata* under irrigated and nonirrigated conditions.

7.3.2   EFFECT ON LAC YIELD

The broodlac yield varied significantly under irrigated and nonirrigated conditions. In this study, all the 60 *semialata* plants were inoculated at 15 g/bush. Broodlac harvest for irrigated plants was 150 g/plant, while it was only 10 g/bush under nonirrigated plants. The broodlac yield ratio was 5:1 and 0.7:1 under these conditions, respectively. It is clear from the data (Table 7.1) that broodlac production could not match even inoculated amount of broodlac, in the case of nonirrigated conditions.

TABLE 7.1   Lac Yield and Yield Ratio of *Flemingia semialata* in Irrigated and Nonirrigated Conditions.

Treatment	No. of plants	Mean weight of inoculated broodlac (g/plant)	Mean broodlac yield (g/plant)	Broodlac yield ratio (output: input)
Irrigated	30	15	150	5: 1
Nonirrigated	30	15	10	0.7: 1

The negligible broodlac production may be attributed to the summer mortality of *kusmi* insects on *semialata* under nonirrigated conditions. The summer sunlight exposure caused desiccation of thinner shoots and consequently death of lac male insects, the time when the insects stop taking plant sap. Better plant growth coupled with enhanced broodlac yield ratio under irrigated condition might be attributed to the sufficient moisture and nutrient supply to the plants.

7.4 SUMMARY

Keeping the root zone of *semialata* plant moist for longer period helps one in sufficient supply of moisture and nutrients to the shoot resulting in good broodlac production. It may be one of the feasible solutions for summer *kusmi* lac crop production on *semialata*.

KEYWORDS

- **broodlac production**
- **lac inoculation**
- **nontoxic resinous product**

REFERENCES

1. Mishra, Y. D. Technology of Lac Cultivation on *Kusum*. In *Recent Advances in Lac Culture*; Indian Lac Research Institute (ILRI): Ranchi, India, 2002; pp 138–147.
2. Singh, B. P.; Mishra, Y. D.; Kumar, K. K.; Agarwal, S. C.; Kumar, P. Integration of Lac Cultivation in Agri-Horticultural System in Chhota Nagpur Region. In *Recent Trends in Horticultural Research*; Central Horticultural Experimental Station: Ranchi, 2000; pp 289–293.
3. Singh, B. P.; Mishra, Y. D.; Kumar, P. Intensive Lac Cultivation for Socio-Economic Upliftment of Lac Growers. In *National Symposium on Lac Industry: Convergence for Resurgence*; Indian Lac Research Institute (ILRI): Namkum, Ranchi, 2004; pp 17–18.
4. Singh, R. K.; Mishra, Y. D.; Baboo, B. Impact of Pitcher Irrigation and Mulching on the Summer Season *(Jethwi)* Lac Crop Sustainability and Pruning Response on Ber (*Ziziphus mauritiana*). *Indian Forestry* **2010,** *136*, 1709–1712.
5. Yadav, S. K.; Mishra, Y. D.; Singh, B. P.; Kumar, P.; Singh, R. K. *Kusmi* Lac Production on *Flemingia semialata*; Technical Bulletin No. 5: Indian Lac Research Institute (ILRI), Namkum, Ranchi, 2005; p 67.

PART II
Irrigation Requirements of Drip Irrigated Crops

CHAPTER 8

ESTIMATION OF CROP WATER REQUIREMENTS

VYAS PANDEY

ABSTRACT

The evaporation and transpiration are two important components in agro-climatic studies and hydrological modeling. The evapotranspiration (ET) from a specific crop is termed as crop evapotranspiration or crop water requirements (ETc), which vary with its growth stages and growing environment. ETc can be measured by lysimeter, but such instruments are not feasible to install in actual fields. Hence various approaches are suggested for direct or indirect estimation of ET, among which climatological approaches are easy to use. Penman–Monteith method is more accurate for estimating potential evapotranspiration (PET) or reference evapotranspiration (ETo). To estimate water requirement of crops, PET is multiplied with the crop coefficient (Kc). Over the years the report from different regions of the world has revealed that ETo is decreased significantly due changes in weather variables. In this chapter, ETo was found to vary with location, day, month, and season. The Kc values were corrected for mid-stage and end-stage of the selected crops. The ETo was estimated for different locations in Gujarat, India. The variation of ETc of 16 different crops including horticultural crops are presented and discussed in this chapter.

8.1 INTRODUCTION

Irrigation water plays a significant role in agricultural sector. Due to uneven distribution of water on the globe, the proper management of water resources is extremely important. There is a great water demand for agriculture, industry, domestic, and all other sectors. Water productivity of irrigated agriculture needs to be augmented. The adoption of exact or correct amount of water and correct timing of application is very essential in irrigation scheduling for optimum crop production. The accurate estimation of irrigation water requirements can save water in agriculture sector. The crop water requirement (ETc) depends on crop growth stage, climatic factors, cropping techniques and patterns, and irrigation methods.

ET varies with variation in climatic parameters, crop types, and vegetation types.[3] To determine ET and ETc, various detailed methods have been given by several scientists.[15] The irrigation scheduling based on ETc can be determined by multiplying crop coefficient (Kc) and reference evapotranspiration (ETo).[10] The ETc depends on age of the crop, crop growing season, its location, and management strategies, and their estimation needs information on reference ETc, Kc, etc. Unavailability of such information may lead to either under or over application of water. Among different methods for estimating ET rates, the climatological based methods are widely used.

This chapter describes various methods used for (1) estimating ETo with its spatial and temporal variation; (2) determining the corrected Kc values. The chapter also describes determination of ETc for different crops under different environmental conditions and micro irrigation systems.

8.2   ESTIMATION OF EVAPOTRANSPIRATION

To estimate actual ET, a term potential evapotranspiration (PET) is defined, which is the maximum ET from a vegetation completely covering the ground surface that has unlimited water supply to its roots.[40] ETo is defined as the water loss through evaporation process to the atmosphere and transpiration (loss of water from a surface of 8–12 cm tall green grass cover, actively growing without any stress with well-watered and completely shading the ground). Although there is a slight difference between the two terms PET and ETo, yet these terms have been used by many researchers for the same purpose.[16,17] The empirical estimation of ETo requires climatic data, such as solar radiation, temperature, wind speed, and relative humidity.

ET can be measured by lysimeter, but such instruments are not feasible to install in the desired fields. Therefore, various approaches are suggested for direct or indirect estimation of ET. The simplest one is the residual of water balance methods, where precipitation/irrigation is taken as input variables and soil moisture storage and run-off/deep percolation as outputs are measured, giving the balance as ET. More advance techniques are also being used, such as micrometeorological, which take into account the instantaneous variations/fluctuations in weather elements (such as Bowen ratio, energy balance,[5] and eddy covariance method[39]). The large aperture scintillometers have also been used to quantify the surface energy fluxes over a given region.[29,38,42]

Such methods give more accurate estimation of evaporation and/or transpiration. However, these cannot be applied in the field for determining ETc and irrigation scheduling. Moreover, these methods require more sophisticated instrumentations. Hence, climatological approaches are widely accepted and used. For estimating PET or ETo, different methods are available. Based on requirement of climatic parameters, these are based on temperature, pan evaporation, radiation, and energy balance–based methods.

8.2.1 POTENTIAL EVAPOTRANSPIRATION METHODS BASED ON TEMPERATURE

The method developed by Thornthwaite is one of the simplest method to compute the PET (ETo)[40]:

$$ETo = 1.6 \left[(10 \ T/I)^a \times (N/12) \times (m/30) \right] \tag{8.1}$$

where a is the constant that varies with heat index (I); I is the $\sum_{i=1}^{12} (T/5)^{1.514}$; N is the daylight hours/maximum possible sunshine hours (h); m is the number of days in a month; T is the mean air temperature (daily mean monthly °C).

Another mathematical relationship for computing ETo was developed using air temperature and daylight factors[4] as given as follows:

$$ETo = k \left[p(0.46 \ T + 8.13) \right] \tag{8.2}$$

where ETo is the reference evapotranspiration (mm) for daily or monthly; k is the monthly consumptive use coefficient (depending upon type of vegetation, time of year, and location); p is the mean monthly percentage of annual day time hours (percentage of total day time hours for period of daily); T is the mean temperature (°C).

The accuracy of ET estimation on daily basis has limitation; therefore, it is suitable for longer time periods.[43] The following equation is used to estimate PET (ETo) using temperature, relative humidity and latitude[13]:

$$ETo = M_f (1.8 \ T + 32) C_H \tag{8.3}$$

where ETo is the reference evapotranspiration (mm/month); M_f is the monthly latitude dependent factor; T is the mean monthly temperature (°C); C_H is the relative humidity coefficient.

8.2.2 POTENTIAL EVAPOTRANSPIRATION METHODS BASED ON PAN EVAPORATION

The following empirical equation is used to estimate ETo using temperature, humidity, wind speed, solar radiation, and pan evaporation[8]:

$$ETo = 0.755 \left[E_{pan} \times C_T \times C_W \times C_H \times C_S \right] \qquad (8.4)$$

where ETo is the reference evapotranspiration (mm/day); E_{pan} is the pan evaporation (mm/day); C_T is the temperature coefficient; C_W is the wind velocity coefficient; C_H is the humidity coefficient; C_S is the solar radiation coefficient.

Doorenbos and Pruitt[10] developed the following equation for estimating daily ETo by using pan coefficient:

$$ETo = K_p \times E_{pan} \qquad (8.5)$$

where ETo is the reference evapotranspiration (mm/day); K_p is the pan coefficient (varies from 0.67 to 0.83); E_{pan} is the pan evaporation (mm/day).

8.2.3 POTENTIAL EVAPOTRANSPIRATION METHODS BASED ON RADIATION

Since solar radiation is the main source of energy for governing physical processes in atmosphere and at the earth surface, therefore, several methods have been developed based on solar radiation and also involving temperature, wind, and other parameters. Mehta[21] developed the following equation to estimate ETo using incoming short-wave radiation R_s for grass lands:

$$ETo = 0.65 \times \left\{ \frac{\Delta}{\Delta + \Upsilon} \right\} \left\{ (R_s - G) \right\} \qquad (8.6)$$

where ETo is the reference evapotranspiration (mm/day); Δ is the saturation vapor pressure (kPa/°C); Υ is the psychometric constant (kPa/°C); R_s is the solar radiation (MJ/m²/day); G is the ground heat flux (MJ/m²/day).

The equation using average daily radiation and temperature to estimate ETo by Turc[41] is expressed as follows:

$$ETo = 0.013 \times \left[(23.88 \times R_s) + 50 \right] \times T \times \left[(T + 15)^{-1} \right] \qquad (8.7)$$

where ETo is the reference evapotranspiration (mm/day); T is the mean air temperature (°C); R_s is the solar radiation (MJ/m²/day).

Turc method is the simplest and accurate of ETo estimation for humid regions.[15] The following equation uses temperature and net radiation data to calculate ETo[32]:

$$\text{ETo} = \left\{ \frac{\alpha\,\Delta}{\lambda\Delta + \Upsilon} \right\} (R_n - G) \qquad (8.8)$$

where ETo is the reference evapotranspiration (mm/day); a is the calibration factor (assuming values of 1.26); Δ is the slope of saturation vapor pressure (kPa/°C); λ is the latent heat of vaporization (J/kg); R_n is the net solar radiation (MJ/m²/day); G is the soil heat flux (MJ/m²/day); Υ is the psychometric constant (kPa/°C).

Equation to compute ETo using air temperature and radiation data[10] is as follows:

$$\text{ETo} = c \left\{ (0.408W \times R_s) \right\} \qquad (8.9)$$

where ETo is the reference evapotranspiration (mm/day); c is a constant that varies with mean relative humidity and daytime wind speed; W is the another constant that varies with temperature and altitude; R_s is the shortwave solar radiation (MJ/m²/day).

Equation 8.9 is used for ETo estimation,[7] which depends on temperature and sunlight hours.

Equation using temperature and extraterrestrial solar radiation data[14] is as follows:

$$\text{ETo} = 0.0023 \times \left[(T + 17.8) \times R_a \right] \times \left[(T_{max} - T_{min})^{0.5} \right] \qquad (8.10)$$

where ETo is the reference evapotranspiration (mm/day); T_{max} is the maximum daily air temperature (°C); T_{min} is the minimum daily air temperature (°C); T is the mean temperature (°C); R_a is the extraterrestrial solar radiation (MJ/m²/day).

8.2.4 METHODS BASED ON ENERGY BALANCE

Method using the concept of energy supply and transport of water vapor from the surface[31] is used to estimate evaporation from surface of water (ETo) as follows:

$$\text{ETo} = \frac{[(\Delta * R_n / \lambda) + \gamma * E_a]}{[\Delta + \gamma]} \qquad (8.11)$$

where ETo is the open water evaporation rate ($kg/m^2/s$); Δ is the slope of saturation vapor pressure (kPa/°C); R_n is the net radiation (W/m^2); G is the heat flux density into the water body (W/m^2); λ is the latent heat of vaporization (J/kg); γ is the psychrometric constant (kPa/°C); E_a is the isothermal evaporation rate ($kg/m^2/s$). The evaporation rate (kg/m^2) is multiplied with 86,400 to convert it into mm/day.

Modified form of eq 8.11 to account it for a grass surface instead of water surface[10] is given as follows:

$$ETo = C\left[\frac{\Delta}{\Delta+\gamma}*R_n +\frac{\Upsilon}{\Delta+\gamma}*(0.27)(1-0.1u_2)(e_s-e_a)\right] \qquad (8.12)$$

where ETo is the reference evapotranspiration (mm/day); Δ is the slope of saturation vapor pressure (kPa/°C); Υ is the psychometric constant (kPa/°C); R_n is the net radiation (mm/day); u_2 is the wind speed at 2 m height; (e_s-e_a) is the difference between saturated vapor pressure and actual vapor pressure of air (kPa); C is the correction or calibration factor.

Canopy resistance (r_c) term is used in Penman method (P–M) to estimate ETo.[27] This method has been recommended by FAO.[31] The FAO P–M equation is given as follows:

$$ETo = \frac{0.408\{\Delta(R_n-G)+[\gamma(900/T_a+273)u_2(e_s-e_a)]\}}{[\Delta+\gamma(1+0.34u_2)]} \qquad (8.13)$$

where ETo is the reference evapotranspiration (mm/day); R_n is the net radiation at the crop surface ($MJ/m^2/day$); G is the soil heat flux density ($MJ/m^2/day$); T_a is the mean daily air temperature (°C); u_2 is the wind speed at 2 m height (m/s); e_s is the saturation vapor pressure (kPa); e_a is the actual vapor pressure (kPa); e_s-e_a is the saturation vapor pressure deficit (kPa), Δ is the slope of saturation vapor pressure curve (kPa/°C); Υ is the psychrometric constant (kPa/°C).

The detailed procedure to compute daily ETo with eq 8.13 has been described by FAO.[10,22,24]

8.3 VARIABILITY AND TRENDS IN REFERENCE EVAPOTRANSPIRATION

It can be concluded from the discussions in this chapter that all methods for estimating the ETo or PET incorporate one or more climatic parameters. These methods are based on the principal that the climatic variables determine the energy required for evaporation to take place and also transportation of water vapor. The FAO P–M method in eq 8.13 consists of two main components: energy balance component and the aerodynamic component. The contribution of each component in total ETo value may vary with climatic regions of the world. The energy balance component contributed 70–75% of ETo in India.[6,34] Due to global warming and climate change, the atmospheric parameters are changing over the period of time. Therefore, ETo is also bound to vary and change with location, season, and time.

8.3.1 SPATIAL VARIABILITY IN REFERENCE EVAPOTRANSPIRATION

The annual ETo computed by FAO P–M method for 18 stations of Gujarat is presented in Figure 8.1. This Figure implies that annual ETo in Gujarat varies from 1912 to 3041 mm. The highest ETo (3041 mm) is at Arnej, which is part of middle Gujarat, followed by 2937 mm at Targadia and 2847 mm at Amreli of Saurashtra region, and 2805 mm at Viramgam of North Gujarat. The lowest value of annual ETo (1912 mm) is observed at Khedbrahma, which is the northern part of Gujarat, followed by 2101 mm at Paria, 2294 mm at Navsari, and 2343 mm at Surat, all from South Gujarat and 2242 mm at Anand in middle Gujarat. The variation in ETo is mainly due to variation in climatic conditions at these stations.

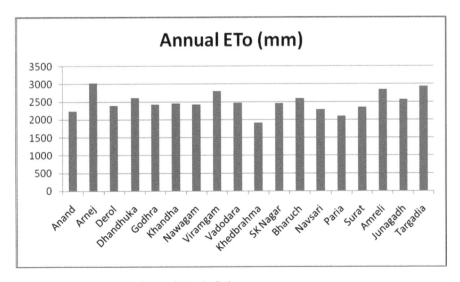

FIGURE 8.1   Variation of annual ETo in Gujarat.

8.3.2   TEMPORAL VARIABILITY IN REFERENCE EVAPOTRANSPIRATION

The monthly ETo at selected stations of Gujarat is presented in Figure 8.2, which indicates variation in ETo varies during different months of the year. ETo is highest during summer season and lowest during winter season. From January to May, the ETo increases continuously, the rate of increase being different in different months with peak ETo in May at all stations. The ETo decreases sharply during June and July due to increase in humidity as a result of onset of monsoon then there was slight decrease during August. It further increases slightly during September and October before reaching to its lowest value during December. These variations are not uniform at all locations, because the energy balance and aerodynamic components of ETo estimation vary with location. Nag et al.[28] have also found that May month had maximum ETo and minimum ETo during December and January for most of the states and the basins of India.

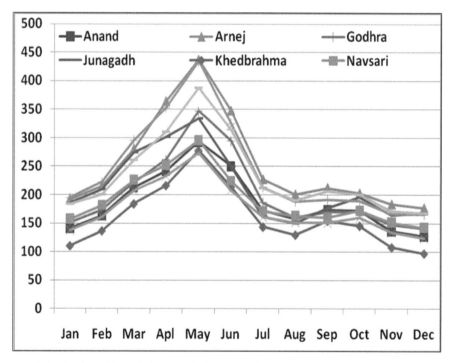

FIGURE 8.2   Monthly variation of ETo (mm) at selected stations of Gujarat.

8.3.3   TRENDS IN EVAPOTRANSPIRATION RATES

Literature review indicates that ET rates are changing during the past several years. In several parts of the world, a decreasing trend in ET has been observed. Under the projected climate change scenarios with increased temperatures, the implications of the ET trends on the hydrological cycle are somewhat controversial. It was found that there is an inverse relationship of temperature with evaporation and ET. With raise in temperature, both evaporation and ET are decreased. Figure 8.3 shows a significant decreasing trend in annual reference ETc at ICRISAT, Patancheru, Hyderabad during the last 35 years (1975–2009).[34,38]

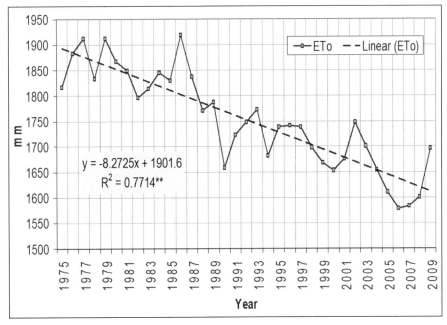

FIGURE 8.3   Trend in reference crop evapotranspiration (ETo) at ICRISAT, Patancheru.[34]

Average annual ETo was decreased by 57 mm per decade or about 3% of the annual total. Changes in the two components (energy balance and aerodynamic) of the ETo were examined to better understand the conspicuous decrease in the ETo and it is seen that the energy balance component showed a positive trend while the aerodynamic component showed a highly significant negative trend (Fig. 8.4). Annual energy component has increased from about 1000 to 1100 mm, while aerodynamic component was decreased from about

800 to 550 mm. Rate of increase for the energy balance component is about 29 mm per decade while the rate of decrease for the aerodynamic component is about 71 mm per decade.

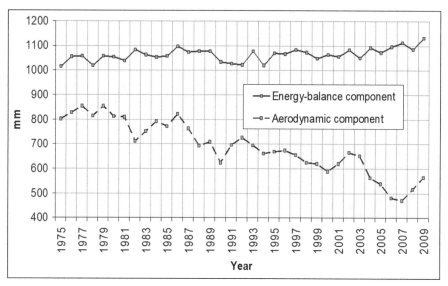

FIGURE 8.4   Trend in energy balance and aerodynamic components of ETo at ICRISAT, Patancheru.

8.4   CROP COEFFICIENTS AND IRRIGATION REQUIREMENTS

The Kc were determined experimentally while measuring ETc using lysimeters.[37] To estimate ETc, Kc and ETo are mostly used.[10]

8.4.1   CROP COEFFICIENTS

The ratio between actual ETc and the ETo is called Kc. Since lysimeters cannot be installed in each field area, it is convenient to apply Kc values of crop to obtain ETc from ETo. The value of Kc changes as the crop grows and crop canopy develops. A schematic diagram of variation of Kc values with crop growth stages is given in Figure 8.5.

The Kc values are less during initial period of crop growth stage when surface of soil is not fully covered by the crop (less than 10% of ground cover). During the development stage of crop till attaining full groundcover, the Kc value increases continuously (Fig. 8.5). It remains more or less constant during the subsequent period till the start of maturity. Thereafter, Kc values are decreased during maturity period till the harvest of the crop.

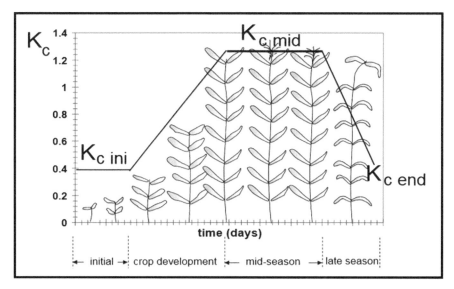

FIGURE 8.5   Variation of Kc values with crop growth stages.

The Kc value also varies with variation in climatic conditions during crop growing season. The Kc are adjusted based on wind and humidity.[10] The Kc (Kc$_{mid}$ and Kc$_{end}$) values for different crops were corrected for climatic conditions of different locations of Gujarat, whereas the Kc values for developmental stage and late-season stage for each of the crop were calculated by linear interpolation[24] as follows:

$$Kc_{mid} = Kc_{mid(tab)} + \left[0.04\left(u_2 - 2\right) - 0.004(RH_{min} - 45)\right]\left[\frac{h}{3}\right]^{0.3} \qquad (8.14)$$

$$Kc_{end} = Kc_{end(tab)} + \left[0.04\left(u_2 - 2\right) - 0.004(RH_{min} - 45)\right]\left[\frac{h}{3}\right]^{0.3} \qquad (8.15)$$

where Kc is the crop coefficient; Kc$_{ini}$ is the crop coefficient at initial growth stage of crop; Kc$_{mid}$ is the crop coefficient at mid-season growth stage of crop; Kc$_{end}$ is the crop coefficient at the end of growing period; Kc$_{mid(Tab)}$ is the value for Kc$_{mid}$ taken from reference table; Kc$_{end(Tab)}$ is the value for Kc$_{end}$ taken from reference table; u_2 is the mean value for daily wind speed at 2 m height over grass during mid-season growth stage (m/s), for 1 m/s $\leq u_2 \leq$ 6 m/s; RH_{min} is the mean value of daily minimum relative humidity during mid-season growth stage (%), for 20% $\leq RH_{min} \leq$ 80%; h is the mean plant height during mid-season growth stage (m) for 0.1 m $< h <$ 10 m.

The corrected Kc values were determined for 11 major crops (such as rice, maize, wheat, pearl millet, Bengal gram, green gram, soybean, castor, groundnut, mustard, and cotton) at different selected locations of Gujarat[30], and these values at different locations are presented in Table 8.1. The Kc_{mid} and Kc_{end} for most of the crops were found to vary across the locations. Generally in dry regions, corrected Kc values are higher than the FAO Kc values.[3,30] Moreover most of the locations of north Saurashtra and Bhal regions recorded higher Kc values.

TABLE 8.1 Corrected Crop Coefficients (Kc) for Different Crops.

Crop	Kc_{ini}	Kc_{mid}	Corrected Kc_{mid}	Kc_{end}	Corrected Kc_{end}	Plant height (m)
Castor	0.35	1.15	1.14–1.17	0.55	0.54–0.57	1.7
Chickpea	0.40	1.00	0.67–1.01	0.35	0.34–0.36	0.4
Cotton	0.35	1.15	1.08–1.15	0.50	0.43–0.50	1.1
Green gram	0.40	1.05	0.96–1.05	0.60	0.51–0.60	0.4
Groundnut	0.40	1.15	1.06–1.15	0.60	0.51–0.60	0.4
Maize	0.30	1.20	1.08–1.20	0.35	0.23–0.35	1.5
Mustard	0.35	1.15	1.13–1.16	0.35	0.33–0.36	1.3
Pearl millet	0.30	1.00	0.87–1.00	0.30	0.18–0.24	1.5
Rice	1.05	1.20	1.11–1.20	0.90	0.90	0.8
Soybean	0.40	1.15	1.05–1.15	0.50	0.41–0.50	0.6
Wheat	0.40	1.15	1.13–1.17	0.41	0.39–0.43	0.8

8.4.1.1 CROP COEFFICIENTS FOR HORTICULTURAL CROPS

A linear relationship was developed between Kc of grape wines grown under lysimeter with leaf area, leaf area index, and percent shaded area.[41] Among these three relationships developed, one with percent shaded area gave the maximum Kc. Such relationships were used to estimate the Kc and water requirement for pomegranate (*Punica granatum* L.) under different management practices in Maharashtra, India.[12,25]

$$Y = \left[(-0.008) + 0.017X\right], R^2 = 0.95** \tag{8.16}$$

where Y is the crop coefficient (Kc); X is the percent shaded area under the deciduous fruit crops.

Equations were developed to determine Kc of several crops including horticultural crops of Maharashtra under micro irrigation system as a function of day and duration of the crop.[11]

8.4.2   CROP WATER REQUIREMENTS

As described earlier, there are different methods to estimate the ETc using climatic factors. To determine actual ETc, reference ETc and Kc are used as follows:

$$ETc = (Kc) \times (ETo) \tag{8.17}$$

where ETc is the actual ETc rate; Kc is the crop coefficient; ETo is the evapotranspiration rate for a grass reference crop.

8.4.2.1   WATER REQUIREMENT UNDER MICRO IRRIGATION SYSTEM

Water requirement of drip irrigated pomegranate is less than the water requirement under surface irrigation.[12,25] The water requirement (W_R) under drip irrigation system was estimated as follows:

$$W_R = [Fa \times ETc] \tag{8.18}$$

where Fa is the area factor (fraction: the ratio of area shaded by the tree and the area occupied by the tree); W_R is the water requirement; ETc is the actual ETc rate.

8.5   CROP WATER REQUIREMENT OF DIFFERENT CROPS

The ETcs are estimated by combining Kc and ETo, both of which vary with the climatic conditions of the area as described in previous sections in this chapter. Moreover, the Kc varies with growth stage, height, and duration of the crop. The ETc also differs with age of the crop, its location and climatic conditions, etc. The ETcs were estimated for crops in India, and these values are presented in this section.

8.5.1   RICE

Rice is grown during *kharif* season in northern part of India. It is also grown during winter and summer season in southern and eastern states of India. ETc of *kharif* rice at Sabour and Patna in Bihar was 546 and 607 mm, respectively.[20] ETc of 640 mm for *kharif* rice and 851 mm for summer rice in Mahi canal command area of Gujarat was determined.[18] Figure 8.6 shows the ETc of *kharif* rice during different growth stages at different stations in Gujarat. The ETc varied between 70 and 95 mm during initial stage, 139–195 mm during developmental stage, 266–315 mm during mid-season stage, and 135–165 mm during late-season with seasonal variation of 618–754 mm in Gujarat.

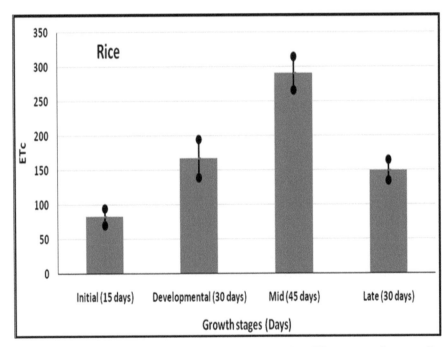

FIGURE 8.6   Crop water requirement of *kharif* rice during different growth stages. Bar shows its variation across the locations in Gujarat.

8.5.2   WHEAT

Wheat is grown during winter (*Rabi*) season in most parts of India. Figure 8.7 shows the ETc of wheat during different growth stages at different locations in Gujarat. ETc varied from 27 to 48 mm during initial stage, 73–135 mm during developmental stage, 201–351 mm during mid-season stage, and 97–156 mm during late-season with seasonal totals of 398–680 mm at locations in Gujarat. Seasonal ETc of wheat at Sabour and Patna is 213 and 243 mm, respectively,[18,20] while it was 565 mm in Mahi canal command area of Gujarat. ETc of wheat in Rajasthan ranged between 173 and 288 mm.[33]

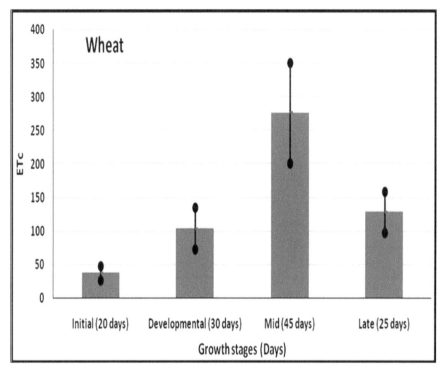

FIGURE 8.7   Water requirement of wheat during different growth stages. Bar shows its variation across the locations in Gujarat.

8.5.3   MAIZE

Maize is grown in all seasons in different parts of India. ETc of maize during three seasons (*kharif, rabi,* and summer) in Bihar was between 292 and 319 mm during *kharif* season, 323 and 372 mm during *rabi* season, and 500 and 591 mm during summer season.[20] In contrast to this, it is reported that the ETc of maize during *rabi* season was slightly less than that of *kharif* season at Anand, Gujarat.[23] Figure 8.8 shows the water requirement (ETc) for *rabi* maize during different growth stages at different stations in Gujarat. ETc was found to vary between 15 and 22 mm during initial stage, 69–110 mm during developmental stage, 174–282 mm during mid-season stage, and 72–107 mm during late-season with seasonal variation of 331–521 mm in Gujarat.

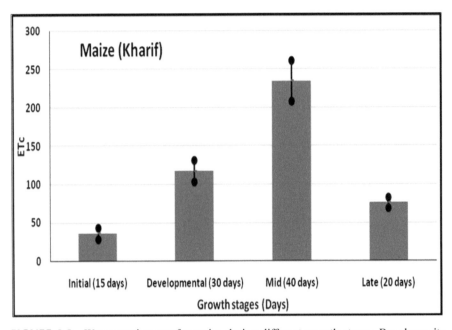

FIGURE 8.8 Water requirement for maize during different growth stages. Bar shows its variation across the locations in Gujarat.

8.5.4   PEARL MILLET

The ETc of *kharif* pearl millet during its different growth stages at different stations in Gujarat are presented in Figure 8.9. ETc was found to vary between 20 and 31 mm during initial stage, 66–93 mm during developmental stage, 162–224 mm during mid-season stage, and 42–57 mm during late-season with seasonal variation of 305–402 mm in Gujarat. Similar range of ETc of *kharif* pearl millet was 25 mm during initial stage, 88 mm during developmental stage, 124 mm at mid-season, and 47 mm at late-season respectively, in central Gujarat.[35] ETc of summer pearl millet (499 mm) was more than that of *kharif* pearl millet (324 mm) at Anand.[23] Even higher ETc of 619 mm for summer pearl millet in Mahi canal command area of Gujarat[18] has been reported while there was a wide variation in ETc of *kharif* pearl millet in western Rajasthan ranging from 308 to 411 mm.[33]

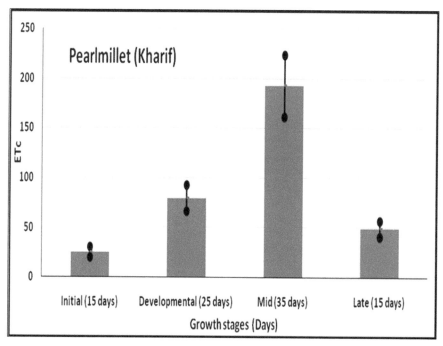

FIGURE 8.9 Crop water requirement of pearl millet during different growth stages. Bar shows its variation across the locations in Gujarat.

8.5.5 SOYBEAN

Figure 8.10 describes ETc of soybean during different growth stages at different locations in Gujarat. ETc was found to vary between 49 and 76 mm during initial stage, 103–148 mm during developmental stage, 236–315 mm during mid-season stage, and 97–131 mm during late-season with seasonal variation of 490–670 mm in Gujarat. ETc of soybean at Anand was 533 mm.[23] In Bangladesh, ETc of soybean was 35, 131, 162, and 51 mm during initial, developmental, mid-season, and late-season, respectively,[26] and these values are lower than that reported in this chapter (Fig. 8.10).

8.5.6 GROUNDNUT

Groundnut is grown both during *kharif* and summer seasons. The ETc of *kharif* groundnut during different growth stages at different stations of Gujarat was presented in Figure 8.11. ETc of *kharif* groundnut was found to vary between 49 and 76 mm during initial stage, 117–172 mm during developmental stage, 213–281 mm during mid-season stage and 105–140 mm during late-season with seasonal variation of 488–670 mm in Gujarat. ETc of summer groundnut (849 mm) was more than that of *kharif* groundnut (536 mm) at Anand.[23] The ETc of *kharif* groundnut at Anantpur, Andhra Pradesh estimated by different methods varied from 281 to 434 mm, which were much more higher than the measured values (455–600 mm).[36]

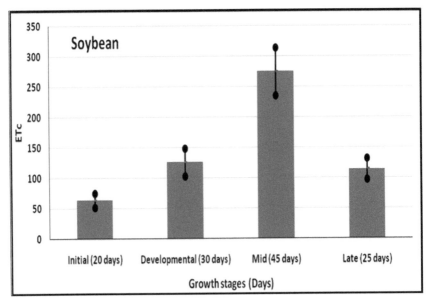

FIGURE 8.10 Crop water requirement during different growth stages of soybean. Bar shows its variation across the locations in Gujarat.

8.5.7 GREEN GRAM

Green gram is also grown during *kharif* and summer season. The ETc of *kharif* green gram during its different growth stages over different stations of Gujarat (Fig. 8.12) is less than that of groundnut (Fig. 8.11). ETc of *kharif* green gram was found to vary between 27 and 40 mm during initial stage, 58–84 mm during developmental stage, 152–190 mm during mid-season stage, and 57–76 mm during late-season with seasonal variation of 299–387 mm in Gujarat. Large variations in ETc of *kharif* green gram in western Rajasthan ranging between 216 and 297 mm have been reported.[33] ETc of summer green gram were 405 mm at Sabour and 476 mm at Patna in Bihar.[20]

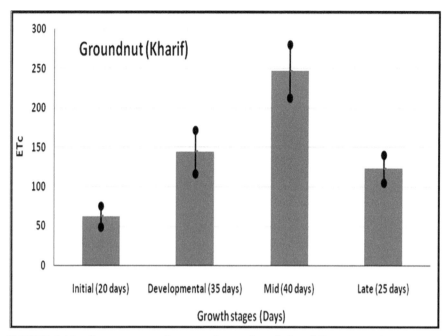

FIGURE 8.11   Crop water requirement of groundnut during different growth stages. Bar shows its variation across the locations in Gujarat.

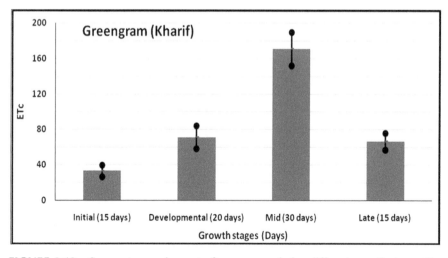

FIGURE 8.12   Crop water requirement of green gram during different growth stages. Bar shows its variation across the locations in Gujarat.

8.5.8 CHICKPEA

Chickpea is grown during winter (*Rabi*) season in most parts of India. Figure 8.13 shows the water requirement (ETc) of chickpea during different growth stages at different stations in Gujarat. ETc of chickpea varied between 23 and 37 mm during initial stage, 79–144 mm during developmental stage, 155–276 mm during mid-season stage, and 77–125 mm during late-season with seasonal totals of 334–581 mm across the locations in Gujarat. In Ethiopia, the ETc of chickpea during initial, developmental, mid-season, and late-season were 37, 114, 205, and 80 mm, respectively,[9] which are similar to the reported values in the present study (Fig. 8.13).

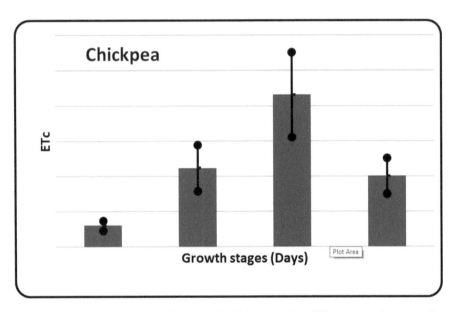

FIGURE 8.13 Crop water requirement of chickpea during different growth stages. Bar shows its variation across the locations in Gujarat.

8.5.9 MUSTARD

Mustard is grown during winter (*rabi*) season in most parts of India. Figure 8.14 shows water requirement (ETc) of mustard during different growth stages and its variation across locations in Gujarat. ETc of mustard varied between 20 and 30 mm during initial stage, 85–143 mm during developmental stage, 178–315 mm during mid-season stage, and 85–139 mm during late-season with seasonal totals of 368–625 mm across the locations in Gujarat. In western Rajasthan, a large variation in ETc of mustard ranging between 214 and 343 mm has been reported,[33] but these values were less than that reported for Gujarat (Fig. 8.14). Similar results were reported in Chhattisgarh, where seasonal ETc of mustard varied between 328 and 373 mm.[19]

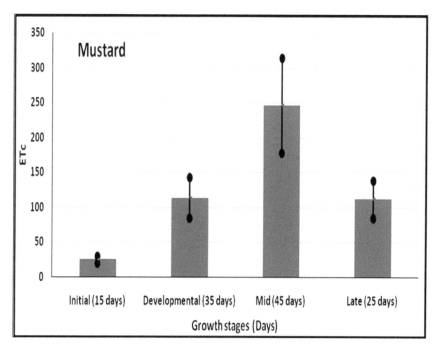

FIGURE 8.14 Crop water requirement of mustard during different growth stages. Bar shows its variation across the locations in Gujarat.

8.5.10 COTTON

Cotton is a medium- to long-duration crop grown during extended *kharif* season in most parts of India. Figure 8.15 shows the ETc of cotton during different growth stages and its variation across the locations in Gujarat. ETc of cotton varied between 59 and 90 mm during initial stage, 137–197 mm during developmental stage, 365–481 mm during mid-season stage, and 141–219 mm during late-season with seasonal totals of 702–986 mm across the locations in Gujarat. The report given by FAO has also states that the water needs for cotton range between 700 and 1300 mm.

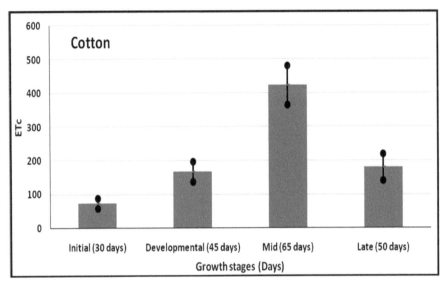

FIGURE 8.15 Crop water requirement of cotton during different growth stages. Bar shows its variation across the locations in Gujarat.

8.5.11 CASTOR

Castor is a long-duration crop grown during extended *kharif* season in most parts of India. Although it is a hardy plant, but due to its long duration, the water requirement is also high. Figure 8.16 shows water requirement (ETc) of castor during different growth stages and its variation across locations in Gujarat. ETc of castor varied between 51 and 68 mm during initial stage, 202–306 mm during developmental stage, 340–487 mm during mid-season stage, and 135–289 mm during late-season with seasonal totals of 760–1148 mm across locations in Gujarat. FAO reports that water need for castor ranged from 700 to 1300 mm.

8.5.12 CLUSTER BEAN

Cluster bean is grown in dry regions of India.[33] The ETc of cluster bean in different districts of western Rajasthan are presented in Figure 8.17. The ETc varied from 244 mm in Pali to 332 mm in Bikaner districts of Rajasthan.

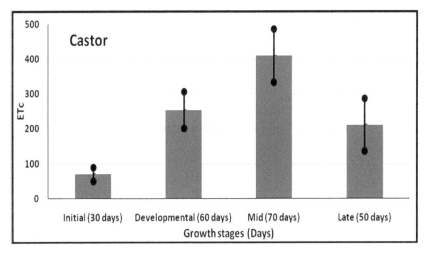

FIGURE 8.16 Crop water requirement during different growth stages of castor. Bar shows its variation across the locations in Gujarat.

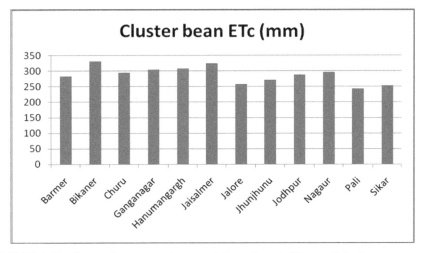

FIGURE 8.17 Seasonal water requirement of cluster bean in Western Rajasthan.

8.5.13 TOBACCO

Tobacco is a medium-duration crop of 120–150 days. The water requirements (ETc) for maximum yield vary from 400 to 600 mm depending on length of growing period and the prevailing climate. The duration of different stages of tobacco are initial stage of 30 days, developmental stage of 40, 50 days under mid-season stage, and late-season of 30 days. ETc of tobacco in central Gujarat was 47, 91, 133, and 43 mm during initial, developmental, mid-season, and late-season stages, respectively with seasonal total of 314 mm.[35]

8.5.14   SAFFRON

Saffron (*Crocus sativus* L.) is a perennial, herbaceous plant that has been cultivated as a spice. The dried stigma of this plant composes the most expensive spice in the world. It is an important cash crop of Jammu and Kashmir State of India. The total water requirement for the saffron was 288 mm, which consists of 80 mm during the initial stage (sprouting to flowering), 134 mm during the mid-season stage (vegetative growth period), and 74 mm during the late-season stage.[2]

8.5.15   APPLE

Apple is a rosaceous fruit tree, grown in states like Jammu and Kashmir, Himachal Pradesh, Uttarakhand, Assam, and Nilgiri Hills. The water requirement of apple in Kashmir Valley during initial was 69 mm, mid-season was 668, and 175 mm at late-season stages, respectively.[1] Figure 8.18 shows the seasonal ETc at locations in Kashmir Valley.

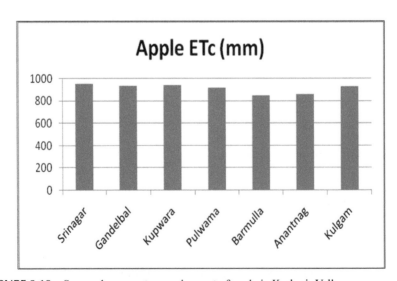

FIGURE 8.18   Seasonal crop water requirement of apple in Kashmir Valley.

8.5.16   POMEGRANATE

Pomegranate, a highly water sensitive crop, is largely grown in marginal lands with the help of fertigation system. The water requirement of pomegranate crop depends on age, season, location, and management strategies.[12,25] The ETc of pomegranate was increased with age of the tree (Fig. 8.19).

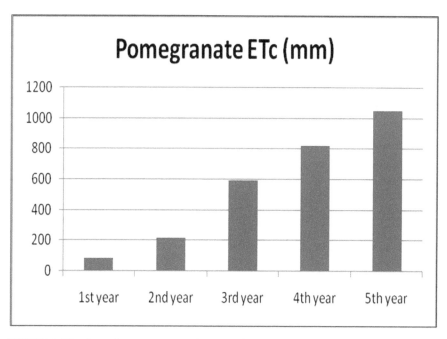

FIGURE 8.19   Annual crop water requirement of pomegranate at different ages of the tree.

The ETc was estimated for several vegetable and horticultural crops[11] in Maharashtra under micro irrigation based on 95% efficiency. The maximum water saving was in pomegranate and lime (88%), followed by papaya (75%) and grapes (73%), while the minimum saving was in summer groundnut (38%) followed by summer onion (40%) and *kharif* groundnut (43%).

8.6   SUMMARY

FAO publication on ETc gives a complete description of methodology. The Kc of different crops are also given in this publication; however, these values need to be corrected for local climatic conditions. ETo estimated for different stations of Gujarat varied temporally and spatially. The energy balance term contributes more to ETo variation in comparison to aerodynamic term. The ETc during different stages for various crops in Gujarat varied with stage and location. The water requirement under micro irrigation system was less than the surface irrigation system.

KEYWORDS

- **crop coefficient (Kc)**
- **crop evapotranspiration**
- **crop water requirement (ETc)**
- **FAO-56**
- **Penman–Monteith method**
- **reference evapotranspiration (ETo)**

REFERENCES

1. Ahmad, L.; Sabah, P.; Saqib, P.; Kanth, R. H. Reference Evapotranspiration and Crop Water Requirement of Apple (*Malus Pumila*) in Kashmir Valley. *J. Agrometeorol.* **2017,** *19* (3), 262–264.
2. Ahmad, L.; Sabah, P.; Saqib, P.; Kanth, R. H. Crop Water Requirement of Saffron (*Crocus sativus*) in Kashmir Kanth Valley. *J. Agrometeorol.* **2017,** *19* (4), 380–384.
3. Allen, R. G.; Pereira, L. S.; Raes, D.; Smith, M. *Crop Evapotranspiration: Guidelines for Computing Crop Water Requirements*; Irrigation and Drainage Paper 56; Food and Agriculture Organization of the United Nations: Rome, 1998; p 304.
4. Blaney, H. F.; Criddle, W. D. *Determining Water Requirements in Irrigated Areas from Climatological and Irrigation Data*; USDA Soil Conservation Service SCS-TP Bulletin 96; Washington, DC, 1950; p 44.
5. Bowen, I. S. The Ratio of Heat Losses by Conduction and by Evaporation from any Water Surface. *Phys. Rev.* **1926,** *27*, 779–787.
6. Chakravarty, R.; Bhan, M.; Rao, A. V. R. K.; Awasthi, M. K. Trends and Variability in Evapotranspiration at Jabalpur, Madhya Pradesh. *J. Agrometeorol.* **2015,** *17* (2), 199–203.
7. Chiew, F. H. S.; Kamaladasa, N. N.; Malano, H. M.; McMahon, T. A. Penman-Monteith: FAO-24 Reference Crop Evapotranspiration and Class-A Pan Data in Australia. *Agric. Water Manage.* **1995,** *28*, 9–21.
8. Christiansen, J. E. Pan Evaporation and Evapotranspiration from Climatic Data. *J. Irrig. Drain.* **1968,** *94*, 243–265.

9. Desta, F.; Bissa, M.; Korbu, L. Crop Water Requirement Determination of Chickpea in Central Vertisol Areas of Ethiopia Using FAO CROPWAT Model. *Afr. J. Agric. Res.* **2015,** *10* (7), 685–689.

10. Doorenbos, J.; Pruitt, W. O. *Crop Water Requirements*; Irrigation and Drainage Paper 24; Food and Agriculture Organization of the United Nations: Rome, 1977; p 156.

11. Gadge, S. B.; Gorantiwar, S. D. Kumar, V.; Kothar, M. Estimation of Crop Water Requirement Based on Penman-Monteith Approach under Micro Irrigation System. *J. Agrometeorol.* **2011,** *13* (1), 58–61.

12. Gorantiwar, S. D.; Meshram, D. T.; Mittal, H. K. Water Requirement of Pomegranate (*Punicagranatum* L.) for Ahmednagar District of Maharashtra State, India. *J. Agrometeorol.* **2011,** *13* (2), 123–127.

13. Hargreaves, G. H. Moisture Availability and Crop Production. *Trans. ASAE* **1975,** *18* (5), 980–984.

14. Hargreaves G. H.; Samani, Z. A. Reference Evapotranspiration Estimations. *Appl. Eng. Agric.* **1985,** *1*, 96–99.

15. Jensen, M. E.; Burman, R. D.; Allen, R. G. *Evapotranspiration and Irrigation Water Requirements*; ASCE Manuals and Reports on Engineering Practices No. 70; ASCE: New York, 1990; p 360.

16. Khandelwal, M. K.; Pandey, V. Estimation of Potential Evapotranspiration by Various Methods in Different Agroclimatic Stations of Gujarat State. *J. Agrometeorol.* **2008,** *10* (2), 439–443.

17. Khandelwal, M. K.; Shekh, A. M.; Pandey, V. Selection of Appropriate Methods for Computation of Potential Evapotranspiration and Assessment of Rainwater Harvesting Potential for Middle Gujarat. *J. Agrometeorol.* **1999,** *1* (2), 163–166.

18. Khandelwal, S. S.; Dhiman, S. D. Irrigation Water Requirements of Different Crops in Limbasi Branch Canal Command Area of Gujarat. *J. Agrometeorol.* **2015,** *17* (1), 114–117.

19. Khavse, R.; Singh, R.; Manikandan, N. Crop Water Requirement and Irrigation Water Requirement of Mustard Crop at Selected Locations of Chhattisgarh State, India. *Ecol. Environ. Conserv.* **2014,** *20* (Suppl.), S209–S211.

20. Kumar, S. Reference Evapotranspiration (ETo) and Irrigation Water Requirement of Different Crops in Bihar. *J. Agrometeorol.* **2017,** *19* (3), 238–241.

21. Makkink, G. F. Testing the Penman Formula by Means of Lysimeters. *J. Inst. Water Eng.* **1957,** *11* (3), 277–288.

22. Mehta, R.; Pandey, V. Reference Evapotranspiration (ETo) and Crop Water Requirement (ETc) of Wheat and Maize in Gujarat. *J. Agrometeorol.* **2015,** *17* (1), 107–113.

23. Mehta, R.; Pandey, V. Crop Water Requirement (ETc) of Different Crops of Middle Gujarat. *J. Agrometeorol.* **2016,** *18* (1), 83–87.

24. Mehta, R.; Pandey, V. *Crop Water Requirement, Evapotranspiration: Estimations*; Unpublished Report by Department of Agricultural Meteorology, Anand Agricultural University (AAU), Anand 388110, Gujarat, India; Write and Print Publications: New Delhi, Feb 22, 2018; p 106. https://www.writeandprint.com/Contact.aspx.

25. Meshram, D. T.; Gorantiwar, S. D.; Sharma J.; Babu, K. D. Influence of Organic Mulches and Irrigation Levels on Growth, Yield and Water Use Efficiency of Pomegranate (*Punicagranatum* L.) *J. Agrometeorol.* **2018,** *20* (3), 196–201.

26. Mila, A. J.; Akanda, A. R.; Sarkar, K. K. Determination of Crop Coefficient Values of Soybean (*Glycine max* [L.] Merrill) by Lysimeter Study. *Agriculturists* **2016,** *14* (2), 14–23.

27. Monteith, J. L. Evaporation and Environment. *Symp. Soc. Exp. Biol.* **1965,** *19,* 205–224.

28. Nag, A.; Adamala, S.; Raghuwanshi, N. S.; Singh, R.; Bandyopadhyay, A. Estimation and Ranking of Reference Evapotranspiration for Different Spatial Scales in India. *J. Indian Water Resour. Soc.* **2014,** *34* (3), 35–45.

29. Nigam R.; Mallick, K.; Bhattacharya, B. K., Pandey, V.; Patel, N. K. Heat Flux Estimation from MODIS TERRA-AQUA and Validation Over a Semi-arid Agroecosystem Using Model Simulation. *J. Agrometeorol.* **2008,** *10* (Special issue), 75–81.

30. Pandey, V.; Mehta, R. *Reference Evapotranspiration and Water Requirement of Crops in Gujarat*; Anand Agricultural University Report AAU/ICAR/Agmet/ES/Report-1; Anand Agricultural University: Anand, India, 2018; p 94.

31. Penman, H. L. Natural Evaporation from Open Water, Bare Soil and Grass. *Proc. Roy. Soc. Lond.* **1948,** *193A,* 120–146.

32. Priestley, C. H. B.; Taylor, R. J. On the Assessment of Surface Heat Flux and Evaporation Using Large Scale Parameters. *Mon. Weather Rev.* **1972,** *100,* 81–92.

33. Rao, A. S.; Poonia, S. Climate Change Impact on Crop Water Requirements in Arid Rajasthan. *J. Agrometeorol.* **2011,** *13* (1), 17–24.

34. Rao, A. V. R. K.; Wani, S. P. Evapotranspiration Paradox at a Semi-arid Location in India. *J. Agrometeorol.* **2011,** *13* (1), 3–8.

35. Rao, B. K.; Kumar, G.; Kurothe, R. S.; Mishra, P. K. Determination of Crop Coefficients and Optimum Irrigation Schedules for Bidi Tobacco and Pearl Millet Crops in Central Gujarat. *J. Agrometeorol.* **2012,** *14* (2), 123–129.

36. Reddy, P. R. Computing Water Requirement of Groundnut. *Ann. Arid Zone* **1988,** *27* (3 & 4), 247–251.

37. Rowshon, M. K.; Amin, M. S. M.; Mojid, M.; Yaji, M. Estimated Evapotranspiration of Rice Based on Pan Evaporation as a Surrogate to Lysimeter Measurement. *Paddy Water Environ.* **2013,** *13* (4), 356–364.

38. Singh, R.; Singh, K.; Bhandarkar, D. M. Estimation of Water Requirement for Soybean (*Glycine Max*) and Wheat (*Triticumaestivum*) in Madhya Pradesh. *Indian J. Agric. Sci.* **2014,** *84* (2), 190–197.

39. Swinbank, W. C. The Measurement of Vertical Transfer of Heat and Water Vapor by Eddies in the Lower Atmosphere. Common Wealth Scientific Research Organization, Australia. *J. Meteorol.* **1951,** *8* (3), 135–145.

40. Thornthwaite, C. W. Approach Towards a Rational Classification of Climate. *Geogr. Rev.* **1948,** *38,* 55–94.

41. Turc, L. *Evaluation des besoinsen eau d'irrigation, evapotranspiration potentielle, formuleclimatiquesimplifiée ET mise a jour* (Assessment of Irrigation Water Requirements, Potential Evapotranspiration, Simplified Climate Formula and Update). *Ann. Agron.* **1961,** *12,* 13–49.

42. Williams, L. E.; Ayars, J. E. Grapevine Water Use and the Crop Coefficient Are Linear Functions of the Shaded Area Measured Beneath the Canopy. *Agric. For. Meteorol.* [Online] **2005,** *132,* 201–211. DOI: 10.1016/j.agrformet.2005.07.010.

43. Wright, J. L. Evapotranspiration and Irrigation Water Requirements. In *Advances in Evapotranspiration; Proceedings National Conference Advances Evapotranspiration*; American Society of Agricultural Engineers, Chicago, IL, 1985; pp 105–113.

CHAPTER 9

POTENTIAL APPLICATIONS OF DSSAT, AQUACROP, APSIM MODELS FOR CROP WATER PRODUCTIVITY AND IRRIGATION SCHEDULING

MUKHTAR AHMED, SHAKEEL AHMAD, SHAH FAHAD, and FAYYAZ-UL-HASSAN

ABSTRACT

This chapter documents the identification of possible water risk indicators and how further generated information can be used by policy-makers to design options for sustainable water management. Pakistan is facing real-time challenges of river basin and freshwater management. Pakistan is mainly dependent on annual influx of Indus River for freshwater availability and is facing massive hydraulic challenges. Annual average rainfall in this region is <240 mm, and per capita (1000 m^3) availability in 2035 projections needs necessary implications and proposed way out. This chapter discusses the concept of virtual water and water footprinting to conserve water on a broader scale. Applications of simulation models (such as DSSAT, AquaCrop, and APSIM) in irrigation scheduling have been considered as management tools to minimize water losses, increasing water use efficiency, and improving productivity of agricultural crops.

9.1 INTRODUCTION

Water being a finite commodity has crucial importance for human consumption, farming, and numerous activities of our daily life. Water quality and availability are critical to human health. These days, water scarcity and water security are challenging issues for policy-makers and educators. According to the latest study, only 2% fresh water is available for human consumption and 98% is salty. Each continent on the Earth is being affected by water scarcity and about 1.2 billion persons have no access to fresh water. Phenomenal converging drivers of its unavailability are a mismatch between availability, demand, living standards, and expansion of irrigated agriculture[79] and depletion of resources.[47] Moreover, essences of absolute and economic water scarcities are due to water resources inadequacy and poor management of existing resources. Hydrological variability due to prevailing management approaches, economic policies, and planning plays key functioning factors for water scarcity.[28]

Water Research Institute (WRI) Aqueduct is measuring, mapping, and understanding water risks around the globe. The global risk mapping tool can help one to understand the water risks and possible opportunities worldwide. These indicators are valuable to have national risk benchmarking and comparison as they have been used by Food and Agriculture Organization (FAO)-AQUASTAT.

Aqueduct water risk indicators are as follows:

- overall water risk (aggregated measures of all indicators);
- physical risk quantity (deals with water quantity, e.g., droughts or floods);
- baseline water stress;
- interannual variability;
- seasonal variability; and
- flood occurrence (Fig. 9.1). The figure indicates the water scenario across the globe and reveals that most of the countries are under severe water stress.[27]

Projected country water stress in 2040 under the business-as-usual scenario depicted that most of the countries will be under severe water stress (Fig. 9.2). Therefore, it is a must that such type of projected information should be used by policy-makers for quantification of climate change and water resources. Ultimately, this will help one to do investments and collaboration in the water sector to design policies for sustainable water management.

Among the South Asian countries, Pakistan and India are facing real-time challenges of river basin and freshwater management. Per capita availability (1200 m³) of fresh water in South Asia compared to other countries (2500–15,000 m³) reveals its vulnerability showing severe water crises.[70] Pakistan is mainly dependent on the annual influx of Indus River for freshwater availability and is facing massive hydraulic challenges. Annual average rainfall in this premise is under 240 mm, and per capita (1000 m³) availability in 2035 projections is an alarming condition, needs necessary implications, and proposed way out.[53] Without knowing water resources, implementation of effective water management strategies is useless.

FIGURE 9.1(a)   Overall **water risk.**

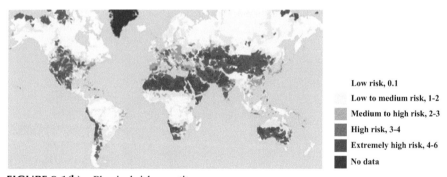

FIGURE 9.1(b)   Physical risk quantity.

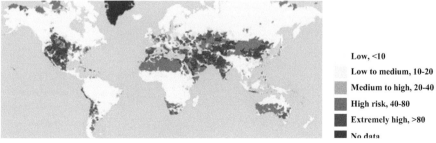

FIGURE 9.1(c)   Baseline water stress (%).

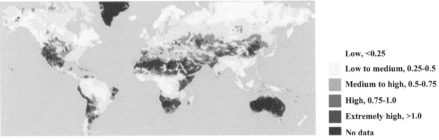

FIGURE 9.1(d)   Inter-annual variability.

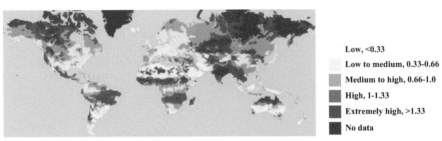

FIGURE 9.1(e)   Seasonal variability.

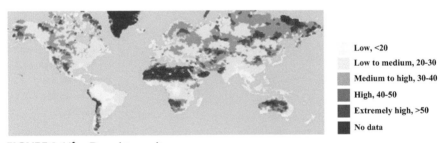

FIGURE 9.1(f)   Drought severity.

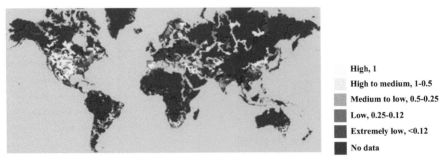

FIGURE 9.1(g)   Upstream storage.

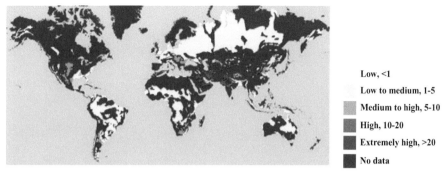

FIGURE 9.1(h)   Groundwater stress.

Source: Modified and compiled with information from Aqueduct Water Risk Atlas, Washington, DC: World Resources Institute, wri.org, https://doi.org/10.46830/writn.18.00146.

Water accounting system proposed by the International Water Management Institute (IWMI) for its productivity, use, and depletion of water resources provides a framework. This methodology is based on mass conservation that inflows must be equal to outflows, including changes in storage, and classifies human interventions to hydrologic cycle (Fig. 9.3).[78] Latest perspectives for water management are virtual water and its footprinting.

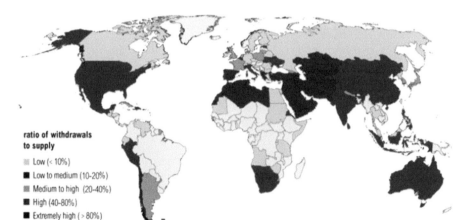

FIGURE 9.2   Global water scarcity projection by 2040.

Source: Water Research Institute.

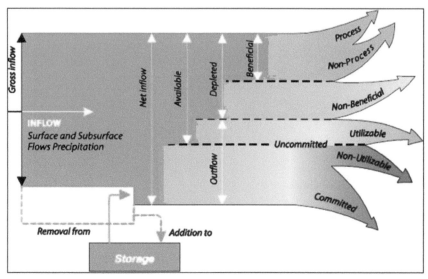

FIGURE 9.3   Chart for water accounting.

Source: Created with information from Refs. [61, 78].

This chapter describes different terminologies related to IS and explores potential and adaptability of simulation models as important techniques to have improved IS in the context of climate change.

9.2   TECHNICAL TERMS FOR IRRIGATION WATER MANAGEMENT

9.2.1   SATURATION (Θ_s)

It is the maximum amount of water that soil can store within it. Total volume equals the water and soil particles with no air spaces left in the soil column. Drainage occurs under the force of gravity until the equilibrium is achieved. When an additional water level rises from the saturation point, then the flooding occurs.

9.2.2   FIELD CAPACITY

When a field is fully irrigated, some water is retained within the soil without drainage due to gravity. When this water remains up to 48 h after saturation, this level of water is called field capacity (FC) of soil. It varies with the soil texture.

9.2.3   PERMANENT WILTING POINT

It is the least amount of water that remains in the soil and is not available to the plants. Before attaining this point, permanent damage starts occurring in the plants.

9.2.4   PLANT AVAILABLE WATER

It is the total amount of water, which is available to plants. This value is obtained by subtracting the value of permanent wilting point (PWP) from the FC.

$$PAW = FC - PWP \tag{9.1}$$

9.2.5   MAXIMUM ALLOWABLE DEPLETION

This is the available portion of water, which is depleted before the need for irrigation. It is considered as the initiating point.

9.2.6   READILY AVAILABLE WATER

It is the easily available portion of water to the plants when they are exposed to minimal stress. IS should be done immediately, when readily available water (RAW) level starts depleting. It can be calculated as follows:

$$RAW = MAD \times PAW \tag{9.2}$$

9.2.7   ROOT ZONE

It is the depth from the soil surface to the deepest layer of the plant. It varies in agronomic and horticultural crops, while in former one root zone is only up to several inches that grows throughout the season to up to the several feet. An amount of water can be calculated by multiplying the root zone depth with the fraction of soil water holding characteristics, which is unitless.

9.2.8   INFILTRATION RATE

It is the speed of water that enters in the soil. It greatly varies with the soil type, amount of water already in the soil, and land slope.

9.2.9   SOIL WATER BALANCE

It is a process of keeping the record of the inputs, outputs, and water storage in the root zone established on the law of mass conservation.

$$\Delta S = I + R - ET_c - P_d - RO + U \qquad (9.3)$$

where ΔS is storage; I is irrigation; R stands for rainfall; ET_c is crop evapotranspiration; P_d is deep percolation; RO is surface runoff; and U is upflux from shallow groundwater table.

9.2.10   EVAPORATION

It is a process in which liquid phase changes into gaseous phase due to an increase in temperature. Or it is a removal of water from surface due to energy sources, such as solar radiation or wind. It is a fundamental part of water cycle and is constantly occurring in nature.

9.2.11   TRANSPIRATION

It is the removal of water from the plant surface. It regulates the temperature of plant by producing the cooling effect.

9.2.12   EVAPOTRANSPIRATION

It is a general term that is used for total water transferred from the root zone by evaporation and transpiration from the plants into the atmosphere.

9.2.13   REFERENCE EVAPOTRANSPIRATION (ET_o)

It is ET_o from the reference surface, which has sufficient amount of water. It is denoted by ET_o. It is a function of weather.

9.2.14   CROP EVAPOTRANSPIRATION (ET_c)

It is specific to the crop, which is considered for irrigation. It is quantified as $[(ET_o) \times (K_c)]$.

9.2.15   CROP COEFFICIENT (K_c)

It is an average ratio of (ET_c) to (ET_o). It varies with plant type, soil type, and weather.

$$ET_c = K_c \times ET_o \qquad (9.4)$$

9.2.16   TOTAL RAINFALL (R)

It is the total amount of precipitation received on a specific area for a given time. It is also measured with a gauge.

9.2.17   NET RAINFALL (R_N)

It is the amount of rainfall that reaches a watercourse canal or the concentration point as a direct surface flow.

9.2.18   EFFECTIVE RAINFALL (R_E)

It is an amount of rainfall, which is incorporated and becomes part of soil water balance. From total daily rainfall, <5 mm in dry spell will not be taken as effective. At FC or above, value of effective rainfall is zero. If the soil is below FC level, then R_E would bring back the level of water in the soil to FC.

9.2.19   CROP WATER REQUIREMENT (CWR)

It is the amount of water needed by plants to grow optimally. This value is the difference between crop evapotranspiration and effective rainfall.

9.2.20   NET IRRIGATION (I_N)

It is the quantity of irrigation that is required to fulfill requirements of plant water.

9.2.21   IRRIGATION EFFICIENCY

It involves the factor that is responsible for uniform and efficiency of an irrigation system. When irrigation efficiency (IE) is low, more water in the form of irrigation will be needed to supply to the plants that can reach to its root zone. Generally for furrow irrigation, IE ranges from 30 to 70%.

9.2.22   GROSS IRRIGATION (I_G)

It is the total amount of water, which is needed so that water can reach down to the root zone. It increases, as IE decreases.

9.2.23   EFFECTIVE IRRIGATION (I_E)

It is the amount of irrigation that reaches down to the root zone. Under ideal scenario, it is equal to the net irrigation.

9.2.24   SURFACE RUNOFF (RO)

It is an amount of water that fails to enter in the soil. It occurs when the irrigation or rainfall rate is greater than infiltration rate of the soil.

9.2.25   DEEP PERCOLATION (P_d)

It is the movement of water outside of the root zone. It can also be referred to as leaching, because it moves deeper into the soil.

9.2.26   UPFLUX (U)

It is an amount of water that moves upward due to shallow ground water table.

9.2.27   GRAVIMETRIC WATER CONTENT

It is the amount of water in the soil and is calculated by measuring the weight of the water compared to the weight of soil. This method involves the collection of samples from required depth and field area, weighing the sample that accounts for fresh weight, oven drying the sample preferably for 24 h at 105°C, and reweighing the sample. This will give the difference, which is the weight of water in the soil. This method gives the gravimetric content of water.

$$\text{Gravimetric water content} = \frac{Weight_{wet} - Weight_{dry}}{Weight_{dry}} \qquad (9.5)$$

9.2.28   VOLUMETRIC WATER CONTENT

It is the measurement of water volume and is compared with total volume of soil, water, and air. This calculation method gives soil water content in the volume of water per volume of the soil column. Another way to determine it is by multiplying the gravimetric with bulk density.

$$\text{Volumetric water content} = \frac{V_{water}}{V_{water} + V_{air} + V_{soil}}$$

$$\text{Volumetric water content} = \text{gravimetric water content} \times \text{bulk density} \qquad (9.6)$$

9.3   VIRTUAL WATER

Usually required virtual water is calculated as follows:

$$VWR[n,c,t] = cp[n,c,t] \times SWD[n,c] \qquad (9.7)$$

where *VWR* is the virtual water requirement (m³/year) of the state n for a crop c in year t; *CP* is the crop production (t/year) of the country n in year t of crop c; and *SWD* is the specific water demand (m³/t) of crop c in state.

Moreover, *specific water demand* is calculated by

$$SWD[n,c] = \frac{CWR[n,c]}{CY[n,c]} \qquad (9.8)$$

where *SWD* is the specific water demand; GSWD is the gross specific water demand; n is the country; c is the crop; *CWR* is the crop water requirement; and *CY* is the crop yield.

9.4   WATER FOOTPRINTING VERSUS IRRIGATION SCHEDULING

Water used through a crop cycle is related to productivity. Water footprinting is a technique to measure relationship between water-use and crop productivity without providing information regarding irrigation management. It is, therefore, necessary that information on water-excess, deficit-irrigation, or water-needed for cropping systems and practices must be added in crop-water footprint analysis at the farm level. The summation of gray, green, and blue components represents crop growth footprints and is expressed as m³/t.

$$WF_c = WF_{green} + WF_{blue} + WF_{gray} \tag{9.9}$$

The green (WF_{green}, m³/t) and blue (WF_{blue}, m³/t) water footprinting components are calculated by dividing CWU (crop-water-use) with crop yield (Y) as follows:

$$WF_{green} = \frac{CWU_{green}}{Y}$$

$$WF_{blue} = \frac{CWU_{blue}}{Y} \tag{9.10}$$

Gray component for any key crop (m³/t) is quantified as the rate of chemicals applied ha^{-1} times to leaching fraction ($\propto$) divided by maximum tolerable concentration (C_{max}) minus natural concentration for pollutant being considered (C_{nat}). Pollutants primarily comprise fertilizers (N, P, etc.) and, to a lesser extent, chemicals like pesticides and herbicides.

$$WF_{gray} = \frac{(\propto \times AR)/(C_{max} - C_{nat})}{Y} \tag{9.11}$$

Green plus blue components in crop-water-use (CWU, m³/ha) are quantified on the basis of accumulated daily ET_c (mm/day) over whole growth cycle. Factor 10 is employed to convert water depths in mm into water volumes per unit surface (m³/ha).

$$CWU_{green} = 10 \times \sum_{d=1}^{\lg p} ET_{cgreen}$$

$$CWU_{green} = 10 \times \sum_{d=1}^{\lg p} ET_{cblue} \tag{9.12}$$

The total water ET from planting to harvest is usually known as green crop water, whereas irrigation and total evpotranspired water are represented by blue crop water use and is cumulative of net water losses and crop evapo-transpired water.

Water requirement of a cultivated crop varies for various agronomic practices, such as seedbed preparation (WAs), irrigation during the crop growth period (WAp), irrigation for the removal of excessive salts from soils (WAss), and frost effects reduction (WAf). It is not necessary that total irrigation volume applied matches with theoretical irrigation needs and are due to inaccurate IS or deficit irrigation (DI). Total irrigation water volume can be assessed by the following equation:

$$CWA_{blue} = WA_g + \sum_{i=1...n} WA_i \qquad (9.13)$$

Timely and accurate delivery of irrigation water to crop plants to maximize productivity by using minimal water is a key concept of IS according to FAO[51] and IWMI.[81] Due to increased competition for crop water needs and dwindling of river basins, alternative strategy (such as IS) is needed rather focusing on current practice systems to cope with emerging challenges.[56]

One of the adverse effects of changing climate due to uncertain rainfall, temperature regimes, and weather extremes is the exacerbated availability, management, and allocation of irrigation water. This sector consumes a major portion of available fresh water worldwide from lakes, rivers, wells, and reservoirs. Irrigated agriculture being more dependent on water resources is considered as highly intensive sector.

Reduced crop productivity due to limited water availability needs advanced precision irrigation technologies to a commensurate increase in crops yield. Dynamic and climate-smart IS for using the appropriate quantity of water according to the type of crop, phenological stage, and weather conditions is a must to enhance commercial crop productivity. Several irrigation methods are available throughout the world according to water prevalence. Besides this, more efficient and effective use of irrigation water will enhance crop yields.[67] Use of less irrigation water for sustainable crop production through advanced irrigation methods and optimization of IS is the need to be developed for potential areas.[6,54].

IS aims to augment the land productivity through increased use efficiency of resources with least environmental externalities and to harness ecosystems for sustainable crop production.[86] Often analogy used for IS is a tipping bucket and estimation of stored water within bucket.[37] Aim is to have more number of crops per unit area on the same piece of land to minimize consequent biodiversity losses and land expansion.[24]

Knowledge of relative root density and distribution is essential for IS in accordance with soil water status.[60] Crop physiological processes and ultimate productivity based upon plant water status and confirms the usefulness of IS.[52] Globally, arid regions being water scarcity areas having lower crop productivity due to less water use efficiency (WUE) can be enhanced through the combination of IS and mulching.[49]

Approaches used for quantitative IS are estimation of soil water balance technique and crop monitoring.[33] IS and management varies with type of soil and is not a simple task in sandy soils due to complex water dynamics. Irrigation system management usually is based on actual field experiences rather on rational basis (soil and agroclimate data) to exacerbate its drawbacks faced by the farming community.[13] Higher crop yield and water saving can be managed through IS in water-limited areas in a given time and space to improve marginal benefits.[64] Similarly, IS is not only significant for water resource management but also dictates for seasonal irrigation events regardless of application efficiency.[23]

Inadequacies observed in current irrigation practices have driven to the development of new tools of IS on the basis of ICT (information and communication technologies).[25] Currently two types of simulation models are used by the scientific community for IS, such as (1) water flux simulation model and (2) soil water balance estimation model.[76] Simulation-based IS can be an alternate appropriate enthusiastic method[88] to maximize WUE, crop yield, and net benefits[21,35]. Usual approaches used for IS are simulation–optimization and optimization–simulation. Moreover, dynamic and nonlinear programming protocols are also used for optimizing IS for the maximum crop productivity.[31]

Linear and nonlinear optimization approaches for irrigation have been used with 10-days interval having limitations of significant variation in soil moisture during simplification of ET and temporal changes in soil water.[19] Water budgeting in the soil is necessary to model IS based on soil water balance, and several simulation models have been developed using this approach.[33] Split application of irrigation water at the needed time not only enhances WUE but also increases crop yield[58] and is of less importance with increase in irrigation frequency.[84]

Several decision support tools have been developed to enhance irrigation use efficiency and crop productivity.[13] Policy making and regional water distributions are planned using water outputs of simulation models. Initially, calibration and testing of these models against real-time data are performed for validation and strengthening user-confidence in model outputs. Simulation and evaluation of modeling software (such as crop and hydrologic models) for different irrigation management strategies are usual practice due to huge investments required for the execution of field-based studies.[5,46] Simulation modeling in combination with long-term climatic variable data and frequency analysis has the potential to schedule timely irrigation for varying environmental conditions.[82]

Crop simulation models (CSMs) are invaluable tools for refining knowledge of agricultural systems functioning.[11,50,63] Genetic algorithms and simulated annealing algorithms are another tools currently used for IS.[1] Devastating integrative evaluation framework for energy and agronomic impacts has been modeled for the assessment of irrigation heterogeneity on soil water management and crop yields.[17]

Allocation of resources[29] and IS[34] based on optimistic simulations has significant advantages for the determination of accurate irrigation.[68] Model-based integration for ET, crop productivity, and algorithm augmentation indicates superiority for appropriate solution. Therefore, simulation approach must be considered to understand the complex relationship of irrigation with ET and crop productivity for IS.[57]

9.5 DECISION SUPPORT SYSTEM FOR AGROTECHNOLOGY TRANSFER: IRRIGATION SCHEDULING

CSMs help for appraisal of crop productivity with weather, soil conditions, and crop practices. Decision support system for agrotechnology transfer (DSSAT) model can be used for agromanagement practices. The scope of DSSAT has now been broadened by linkage with geographical information system. Crop models help one to quantify yield gaps and assist in developing strategies to overcome shortage of inputs limiting crops yield. Increase in competition for water resources has been developed due to shortcomings. Considering this scenario, best water management practices must be followed for efficient utilization of available water.[4]

Several strategies can be considered on the basis of proper application in specific environmental conditions and irrigation systems. Furthermore, water conservation can be enhanced by knowledge-based strategies accompanied by IS. Field experimentation in combination with long-term simulations studies can be helpful in better irrigation management under seasonal variations.

Crop irrigation water optimization over long periods under arid and semi-arid conditions with agriculture system models (such as DSSAT-CERES, APSIM, and AquaCrop) is being used globally for quantification of crop yield.[74] The CERES-Wheat model was calibrated to evaluate the productivity of winter wheat and WUE under three to four irrigation treatments.[7] The results indicated that model simulated the highest yield, when 70 mm of irrigation was applied at jointing. And limited irrigation of 60–90 mm was more profitable in terms of water costs.

TABLE 9.1 Performance of DSSAT Models on Soil and Plant Variables Simulated for Various Irrigation Treatments.

Irrigation treatment	Variables	Performance	Results	Refs.
Drip irrigation with splits at different stages (crown-root initiation, tillering, booting, milking, and dough): 50–100 mm and flooding.	Grain-yield and water productivity of wheat	$R^2 = 0.92$	In contrast to flood method, drip saved 33.0% water with higher grain-yield of 8.3 and 9.3% of wheat water productivity.	[73]
Deficit irrigation: 140 and 100 mm; full irrigation: 400 mm.	Grain filling (wheat)		Higher water use efficiency was recorded with deficit-irrigation at grain-filling stage compared to full irrigation, which requires management of timing of irrigation.	[8]
Moderate irrigation with different timings.	Silking	Difference between observed and simulated value for silking was 0–14 days	Yield was improved by about 30% compared to irrigation at vegetative stage.	[12,66]
Irrigation with water gradient.	ET (maize crop)	Percent error of 2.3–7.64	ET was decreased due to water deficit resulting in 8% yield reduction when stress was induced at vegetative phase; and 50% reduction in yield occurred when reproductive phase of crop was subjected to stress.	[66]
Well irrigated.	Soil water content	RMSE: 4, 35.5%	Soil moisture tension was optimum, but drainage was increased thus reducing crop water use efficiency.	[83]
Well irrigated.	Shoot/leaf nitrogen content (maize crop)	Index of agreement: 0.978	Nutrient use efficiency was enhanced by ET and ultimately resulting in increased nitrogen uptake by plants.	[18,59]
Well irrigated.	Silking	Difference between observed and simulated value was 0–14 days	Provision of water during critical growth stages resulted in yield improvement and vice versa.	[90]
Deficit irrigation: 100 mm.	Jointing		Increased biomass yield.	[8]

TABLE 9.1 *(Continued)*

Irrigation treatment	Variables	Performance	Results	Refs.
Irrigation with varying amount of water and fertilizer gradient.	Uptake of nitrogen (maize)	Difference between observed and simulated value for nitrogen uptake was 70 kg/ha	Interactive effect of nitrogen and irrigation resulted in more nitrogen uptake by crop plants.	[41]
Well irrigated.	Above ground biomass (maize)	Percent error was <25%	Full water provision boosted biomass by 40% and grain yield 25%.	[22]
Five irrigations each of 80 mm at different stages (emergence, jointing, tassel, grain filling, and maturity)	Jointing and tassel (maize)	RMSE: 6%	Yield was increased with irrigation at jointing and tasseling beyond that no significant increase was observed.	[10]
Drip irrigation with splits at different stages (crown-root-initiation, tillering, booting, milking, and dough): 150–400 mm and flooding.	Crop ET	$R^2 = 0.94$	Higher ET was observed with the highest amount of irrigation though drip irrigation.	[73]
Irrigation of 240–400 mm (anthesis); 100 and 140 mm at jointing; one full irrigation (400 mm).	Biomass yield	$R^2 = 0.92$	Biomass yield was same with irrigation at jointing and anthesis compared with full irrigation.	[8]

Different irrigation management strategies were evaluated using CERES-Maize cropping system model to identify economic and less time-consuming method in water-limited scenarios. It was revealed that these computer-based programs are useful tools to manage IS.[4] Under limited water situations, CERES-Maize (DSSAT) model is adequate for large-scale simulation on spatial and temporal basis giving higher yield and net benefits, whereas it was inadequate for individual site having deficit yield conditions.[40] Long-term climatic data (1912–2005) and CERES-Maize were used to allocate most useful irrigation timing between vegetative and reproductive phases under variable environmental conditions for local farming community of Akron—United States.[5]

DSSAT-Maize model was assessed for different irrigation intervals for various crop growth stages, attributes, and crop yield of maize.[10] Outcomes indicated that irrigation variance output depended on the climate scenario. Generally, 1000, 4200, and 4800 m^3/ha irrigation regimes are required for wet, normal, and dry years, respectively. It was inferred that overall efficient utilization of water resources can be assured by reducing water losses to half accompanied by better management practices.

DSSAT-CERES model was also calibrated for irrigation management and WUE in wheat crop.[8] Results for long-run simulation (1980–2012) revealed that grain yield can be increased by 35% if single irrigation of 100 mm is applied at jointing. Irrigation of 140 mm at anther formation increased the grain yield by 68% compared to dryland. Aforementioned results depicted needs of IS and timing with respect to water productivity and efficient utilization.

DSSAT model was used to evaluate crop yield plus water productivity of water-limited cropping systems. The results indicated that model adequately reproduced the phenology, biomass, and yield of wheat, corn, and sorghum. The performance and application of DSSAT family models are summarized in Table 9.1 for different plant and soil parameters to elaborate the efficiency for irrigation management and to provide the policy-makers a pipeline for making strategies to combat water shortages in the present scenario.

Overall summary of the discussions indicated that calibrated model simulated biomass yield efficiently for different irrigation treatments (T1, T2, T6, and T8) in growing season (Fig. 9.4). Overall dryland recorded less biomass yield. Higher WUE was observed when double-irrigation (T6 and T7) along with full irrigation treatment was applied, while lower WUE was observed with T2 and T3 (jointing and booting) under low and high precipitation seasons as shown in Figure 9.4. Double-irrigation at jointing

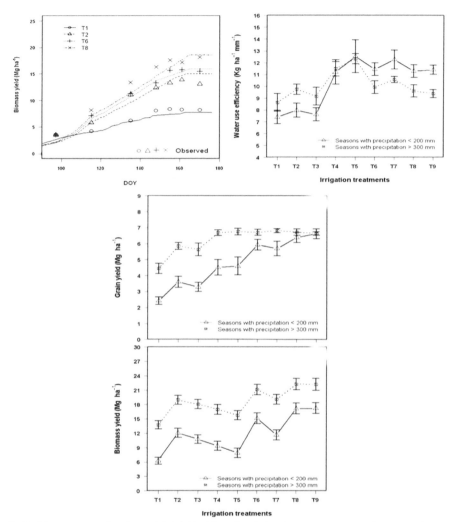

FIGURE 9.4 Application of DSSAT-CSM for irrigation management for a semiarid climate.[26]

and flowering resulted in higher grain compared to full irrigation. Single irrigation (100 mm at booting or 140 mm at flowering) resulted in significant grain yield increment. Simulation for 33 years period depicted variation with time and amount of irrigation. Although response of both grain and biomass yield differed to single irrigation treatment, yet biomass was declined when precipitation fell down average level.

9.6 APPLICATION OF FAO AQUACROP FOR IRRIGATION SCHEDULING

The FAO AquaCrop model was developed to address food security and to evaluate effects of environment along with management on the crop productivity. It can simulate yield response of crop under different water treatments. Further details are available at "www.fao.org/aquacrop." This model has been used by different researchers to study the impacts of irrigation on different crops.[14,15, 36,39,55, 62,72,77,80] Two-years data of rain-fed barley were used to evaluate the AquaCrop model under different treatments (rain-fed, one irrigation at sowing, along with one irrigation at spring). Results indicated that the AquaCrop is a good strategy to simulate barley productivity under different treatments.[2]

The AquaCrop model was also used to simulate potato productivity along with soil moisture at varying water stress situations. Treatments were fully irrigated (I_f), deficit-irrigated (I_d), and not irrigated (I_0) during the experimental years. Results exhibited that the AquaCrop could simulate soil-water-content, canopy cover, and potato productivity with good accuracy.[44] The AquaCrop model was evaluated for potatoes under the following treatments (120, 100, 80, and 60% of water requirement). Field and model results indicated that 80 and 60% treatments were most efficient in the water use.[30]

The AquaCrop model was also calibrated and validated at Bangladesh for rice crop at variable water regimes (continuous standing water, along with irrigation at 3 or 5 days after water disappearance from field) as potential adaptation strategy for water saving. Results showed that the AquaCrop model could predict rice growth and yield accurately.[42]

Cabbage yield and irrigation water use could be optimized by the AquaCrop as it is a useful tool to estimate plant density and yield under different irrigation treatments.[43] DI was evaluated by using the AquaCrop. The results showed that DI approach could be used to maximize crop production. The procedure for DI scheduling was designed with an example of quinoa[5] (Fig. 9.5). Further applications of the AquaCrop model are presented in Table 9.2.

TABLE 9.2 Performance of the AquaCrop Model on Soil and Plant Variables Simulated for Various Irrigation Treatments.

Irrigation treatments	Dependent variable	Performance	Results
Irrigation (0–600 mm) during the season.	Yield	Percent error for cotton, maize, potato, and sunflower is 0.38, 0.25, 0.43, and 0.54, respectively.	Yield was improved by about 11%. The difference between observed and simulated yield for cotton, maize, potato, and sunflower is 0.43, 0.04, 0.55, and 3.8, respectively.[3]
Irrigation levels, that is, 60, 80, and 100%.	Grain yield of wheat cultivars	Percent error for grain yield is 3.48 for calibration and 2.87 for validation.	The maximum water potential was 1.54 kg/m^3 at 60% of ET.[16]
Four irrigations ranging from 200 to 300 mm round the year.	Soil moisture in root zone.	Model accurately simulated the soil moisture in root zone; GY and TDM with present error of <10%.	The maximum grain yield was obtained by scheduling each irrigation at sowing, vegetative growth, and reproductive stage of crop.[77]
Treatment between 3 and 6 for mild water stress and full irrigation.	Evapotranspiration (ET)	The model efficiency (ME) was 0.47.	The percent error for grain yield is 0.09.[87]
Four irrigations: W1: no irrigation; W2: 50% of FC; W3: 75% FC; and W4: full irrigation.	Canopy cover	Model efficiency (ME) for water productivity, biomass, and yield is 0.66, 0.95, and 0.99, respectively.	The comparison between observed and simulated GY, biomass, % water productivity is 0.09, 0.04, and 0.9, respectively.[85]
23 irrigation treatments.	Seed production	Percent error is 19.3 for canopy cover; 19.1 for biomass; and 15.2 for soil moisture.	Full irrigation; 03 times irrigation at vegetative stage; 02 times at reproductive stage; 03 times at reproductive stage; once at vegetative stage.[55]
Three irrigations: full (100%), over (115 and 145% of full); deficit (50–90% of full).	Soil salinity	Model gives satisfactory results at 80% irrigation for salinity.	The maximum WUE at 358–457 and 406–462 mm for silty loam and sandy loam, respectively.[62]
Suboptimal irrigation schedules were calculated on every fourth days.	Weather forecast	The AquaCrop predicts average rainfall for 2009: 50% rain before anthesis and less than half of rain after the flowering stage.	If historical data are used for forecasting, the deviation of results from optimum was <35 mm for irrigation and 75 kg/ha for yield.[39]

TABLE 9.2 (Continued)

Irrigation treatments	Dependent variable	Performance	Results
Three irrigation intensities: full (I_f), deficit (I_d), and no irrigation (I_0).	SWC (soil water content)	The percent error for SWC, TDM production during crop developmental stages, and tuber-yield production for all treatments was 0.164, 0.157, and 0.046, respectively. Model efficiency was 0.973 for 2013 and 0.916 for 2015.	The comparison between observed and simulated SWC at I_f, I_d, and I_0 was 0.381, 0.444, and 0.485.[44]
Irrigation management through threshold of water depletion in root zone, that is, $D_{r,threshold}$	Actual ET, ET_{cact}	Early sowing is far better than late sowing for water saving and more yield production. The value of $D_{r,threshold}$ is 0.6.	The difference between observed and simulated ET_{cact} was —0.23 mm/day.[14]
Irrigation treatments were rain-fed, 01 irrigation at the time of sowing; 01 irrigation in spring.	Percentage of canopy cover (% cc)	Model efficiency for % cc, SWC, and grain yield was 0.98, 0.8, and 0.91, respectively.	Percent error between observed and calculated % cc, SWC, and yield is 8.7, 12.4, and 9.2%, respectively.[2]
Two irrigations: One is deficit; full irrigation.	ET; mean hydraulic conductivity of 500 mm/day on sandy soil.	The error was 0.814 mm/day. Overall the percent error was 25.6% for 2011–2015.	The AquaCrop has not given the satisfactory results for ET, because it considered the canopy cover resulting in low accuracy.[65]

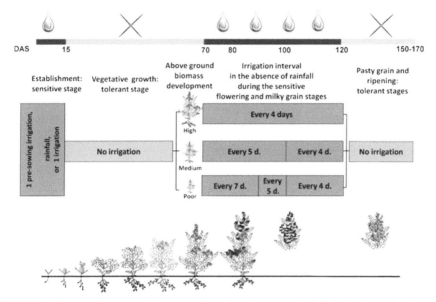

FIGURE 9.5 Application of the AquaCrop for designing deficit irrigation of quinoa. (Created by authors with data from field experiment and the information from Ref. [5].)

9.7 APPLICATIONS OF AGRICULTURAL PRODUCTION SYSTEMS SIMULATOR FOR IRRIGATION SCHEDULING

Irrigation management, cropping system, and environmental strategies can be simulated along with effects of such variations on model's performance. Knowledge gained in designing a model can be of utmost importance to propose improvements in agricultural systems. By altering inputs for simulation and observing outputs, valuable understanding may be achieved for future prediction of Agri-systems. Simulation can be used to develop strategies and policy making considering the outputs. It can also yield results in experimentation with new designs, cropping systems, and IS as a stepping stone toward the implementation and improvement in already developed strategies.

IS management and climate variability effects on soil, crop, and atmosphere system can be assessed through cropping system models. The same models can be used to generalize these results for other climates and sites with some variation in space and time. Although very limited data have been published regarding crop models' performance in irrigation management, yet the present literature depicts that Agricultural Production Systems Simulator (APSIM) model could allow to simulate crop performance in response

TABLE 9.3 Applications of APSIM to Simulate Water Regimes.

Irrigation treatments	Dependent variable	Performance	Results
Conventional flooding (CF); alternate wetting and drying (2, 5, and 8 AWD); with different number of irrigations (2, 3, and 5); and irrigation quantity (835–1728 mm).	Grain yield and irrigation water productivity in rice–wheat system.	RMSE was in acceptable range (24 and 33%) for rice and wheat.	Although more water productivity was observed in conventional flooding, yet at the cost of drastic yield losses. Overall combination of five AWD with five irrigations resulted in more water productivity with grain yield to acceptable level.[32]
Irrigation management with and without mulch in wheat crop.	Grain yield and total biomass.	Under normal water conditions: model did predicted the yield well with R^2 = 0.91 and 0.81 with and without mulch; but underpredicted in water deficit scenario.	Regarding strategic irrigation management, model needs improvement to simulate crop performance better even under water deficit to lower irrigation requirement while sustaining yield.[71]
No irrigation (Totally rain-fed); full irrigation (irrigated wheat crop during all growth stages).	Green leaf biomass and grain yield.	Green leaf biomass was underestimated while grain yield was overestimated when wheat was subjected to stress at growth stage 2.	Stress at growth stage 3 was not detected by model, which therefore over-predicted grain yield.[20]
Rain-fed (642 mm); surface irrigation (195 mm).	Flowering time.	Model predicted flowering time in general but failed to predict flowering time under moisture stress.	Reliable simulation of model can drive more applications in sustainable agricultural system.[89]
Irrigations were applied after emergence based on plant available water (PAW): 110, 109, and 93 mm at sowing of maize.	Grain yield.	Model predicted grain yield satisfactorily under normal irrigations but overpredicted up to 20% in delayed irrigation. R^2 = 0.65 was obtained for simulated and observed values.	APSIM can predict grain yield of maize with varied irrigations but unable to estimate grain yield under drought conditions.[48]

to irrigation management for separate seasons or cropping sequences.[38,71] APSIM is a highly advanced simulator (Table 9.3).

It can be used for multipurpose agromanagement practices, including plant, water, climate, and soil. New versions of APSIM (version 7.× or APSIM Next generation) have been developed to advance and ensure accurate simulation for the development of strategies. The maintenance and development of which is supported by monitoring and software engineering standards. Initiative has been taken to develop and promote APSIM for further use as science models in predicting future scenario. APSIM model was calibrated for the evaluation of water application (irrigation and rainfall) effects on water productivity and dry matter yield of forage crop.[69]

Results indicated that although APSIM predicted dry matter yield and water productivity, yet model was weak to predict seasonal dry matter yield and water productivity at field level. Overall improvement in APSIM is needed to predict accurately every aspect of crop management.

A simulation was carried for canola yield using APSIM in China under various irrigation treatments.[75] It was concluded that potential yield up to 80% can be obtained in wet, medium, and dry seasons using mean irrigation of 330.1, 302.9, and 265.2 mm.

APSIM model was calibrated to evaluate effects of irrigation management on the performance of wheat crop and WUE.[71] The model was used for the prediction of crop growth, yield along with water productivity. Underprediction of model in yield by 600–1000 kg/ha was observed under water deficit conditions. It was concluded that APSIM can be used for modeling wheat under DI, but improvement is needed in the model to simulate crop performance better under irrigation deficit in spite of maintaining yield to minimize water input.

A research trial was conducted in India to assess water balance and crop productivity in response to water management strategies in rice crop (dry seeded rice along with puddled transplanted rice).[45] Different IS (daily-irrigation to alternate-wetting-drying [reducing water application with reduction in yield]) was based on soil water potential (20, 40, and 70 kPa at soil depth of 18–20 cm). It was inferred that DSR 20 kPa resulted in 30–50% water saving due to less drainage and surface runoff compared to puddled transplanted rice with 20 kPa.

Single irrigation increased yield twofold (2.5 t/ha) compared to dryland crops. Some crops showed 30% response to irrigation.[9] APSIM could predict water productivity better on an annual basis compared to a seasonal basis and it indicates that model is good for short-run simulation instead of long run (Fig. 9.6).[69]

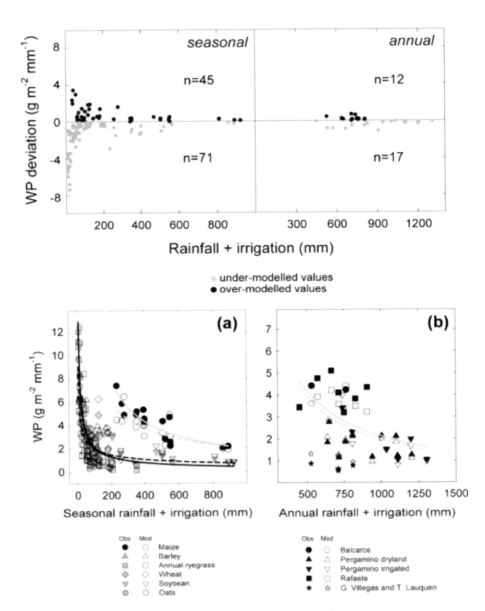

FIGURE 9.6   APSIM application to depict water productivity.[69]

Source: Modified from Ref. [69].

9.8   CHALLENGES AND OPPORTUNITIES

Water is important for agricultural productivity, food security to achieve Sustainable Development Goals. Climate change will have a substantial impact on agriculture-sector by increasing water demand, limiting crop-water-productivity along with reducing water availability across the globe. Therefore, it is essential to develop appropriate policies and strategies by the use of modeling techniques (APSIM, AquaCrop, and DSSAT), which will assist in the mitigation of climate change. This could be as follows:

- Understanding of adaptive pathways and physiological mechanisms of crop responses to water deficits.
- Sustainable improvement in agricultural water productivity.
- Climate-proofing irrigation service.
- Preventing and controlling water pollution.
- Planning, designing besides managing for multiple uses of water resources.
- Using alternative sources of water whereas ensuring food security.
- Promoting   territorial/landscape/ecosystem-based   management approaches.
- Ensuring more efficient usage of water along with food-value-chains.
- Reducing food losses and waste to alleviate pressure on natural resources.
- Diversifying production systems along with income prospects for smallholders.
- Using agricultural trade.

9.9   SUMMARY

FAO estimates that crop production should be increased by 140% to fulfill the demands of the growing population. However, the challenging issue is the unavailability of irrigation water as most of the developing counties are below the "water scarcity" level. The chapter proposes that water-shortage issues can be efficiently solved by adopting innovative technologies to attain water saving without limiting future-agricultural-production. Water saving systems need to be developed by (1) increased investment in water-saving technologies; (2) adoption of an improved engineering system (building of an effective-water-transport-system, efficient water-measuring instrumentation, and installing drip irrigation besides subsurface-drip-irrigation); (3) the development of water-resource management system; and (4) reinforcement of research and development.

We need water saving society and system, which should focus on water transport efficiency, water distribution efficiency, cropping efficiency, and field application efficiency. Simulation modeling is effective to increase WUE in crops. CSM helps one to estimate crop productivity based on weather, soil conditions, and crop management practices. The DSSAT, AquaCrop, and APSIM models have been used for irrigation management and higher water use efficiencies.

ACKNOWLEDGMENT

This study was financially supported by Higher Education Commission (HEC), Pakistan through National Research Program for Universities (NRPU) under project # 6132.

KEYWORDS

- **APSIM**
- **AquaCrop**
- **DSSAT**
- **Indus River**
- **South Asia**
- **sustainable water management**

REFERENCES

1. Abedinpour, M.; Sarangi, A.; Rajput, T. B. S. Performance Evaluation of AquaCrop Model for Maize Crop in a Semi-arid Environment. *Agric. Water Manage.*, **2012**, *110*, 55–66. doi: https://doi.org/10.1016/j.agwat.2012.04.001.
2. Ahuja, L. R.; Ma, L.; Lascano, R. J.; Saseendran, S. A.; Fang, Q. X.; Nielsen, D. C.; Wang, E.; Colaizzi, P. D. *Syntheses of the Current Model Applications Advances in Agricultural Systems Modeling*, 2014, vol. 5; pp 399–438. DOI: 10.2134/advagricsystmodel5.c15
3. Akhtar, F.; Tischbein, B.; Awan, U. K. Optimizing Deficit Irrigation Scheduling under Shallow Groundwater Conditions in Lower Reaches of Amu Darya River Basin. *Water Res. Manage.* **2013**, *27* (8), 3165–3178.
4. Allam, A.; Tawfik, A.; Yoshimura, C. Simulation-based Optimization Framework for Reuse of Agricultural Drainage Water in Irrigation. *J. Environ. Manage.* **2016**, *172*, 82–96.
5. Andarzian, B.; Bannayan, M. Validation and Testing of the AquaCrop Model under Full and Deficit Irrigated Wheat Production in Iran. *Agric. Water Manage.* **2011**, *100* (1), 1–8; doi: http://dx.doi.org/10.1016/j.agwat.2011.08.023.

6. Anothai, J.; Soler, C. M. Evaluation of Two Evapotranspiration Approaches Simulated with the CSM–CERES–Maize Model under Different Irrigation Strategies and the Impact on Maize Growth, Development and Soil Moisture Content for Semi-arid Conditions. *Agric. For. Meteorol.* **2013,** *176*, 64–76; doi: https://doi.org/10.1016/j. agrformet.2013.03.001.

7. Araya, A.; Habtu, S.; Hadgu, K. M. Test of AquaCrop Model in Simulating Biomass and Yield of Water Deficient and Irrigated Barley (*Hordeum vulgare*). *Agric. Water Manage.* **2010,** *97* (11), 1838–1846; doi: https://doi.org/10.1016/j.agwat.2010.06.021.

8. Araya, A.; Kisekka, I. Evaluation of Water-limited Cropping Systems in a Semi-arid Climate Using DSSAT-CSM. *Agric. Syst.* **2017,** *150*, 86–98. doi: https://doi.org/10.1016/j.agsy.2016.10.007.

9. Attia, A.; Rajan, N.; Xue, Q. Application of DSSAT-CERES-Wheat Model to Simulate Winter Wheat Response to Irrigation Management in the Texas High Plains. *Agric. Water Manage.* **2016,** *165*, 50–60; doi: https://doi.org/10.1016/j.agwat.2015.11.002.

10. Balwinder, S.; Gaydon, D. S. The Effects of Mulch and Irrigation Management on Wheat in Punjab, India: Evaluation of the APSIM Model. *Field Crops Res.* **2011,** *124* (1), 1–13; doi: https://doi.org/10.1016/j.fcr.2011.04.016.

11. Ban, H. Y.; Sim, D.; Lee, K. J. Evaluating Maize Growth Models CERES-maize and IXIM-maize under Elevated Temperature Conditions. *J. Crop Sci. Biotechnol.* **2015,** *18* (4), 265–272; doi: 10.1007/s12892-015-0071-3.

12. Bello, Z. A.; Walker, S. Evaluating AquaCrop Model for Simulating Production of Amaranthus (*Amaranthus cruentus*) a Leafy Vegetable, under Irrigation and Rain-fed Conditions. *Agric. For. Meteorol.* **2017,** *247*, 300–310; doi: https://doi.org/10.1016/j. agrformet.2017.08.003.

13. Ben-Nouna, B.; Katerji, N. Using the CERES-Maize Model in a Semi-arid Mediterranean Environment. Evaluation of Model Performance. *Eur. J. Agron.* **2000,** *13* (4), 309–322; doi: http://dx.doi.org/10.1016/S1161-0301(00)00063-0.

14. Briscoe, J.; Qamar, U. *Pakistan's Water Economy: Running Dry*; Oxford University Press: Karachi, Pakistan, 2006, p 89.

15. Carberry, P. *Study to Assess the Technical and Economic Feasibility of Wheat Production in Southern Bangladesh*; Australian Centre for International Agricultural Research: Canberra, Australia, 2008; p 124.

16. Casadesús, J.; Mata, M. General Algorithm for Automated Scheduling of Drip Irrigation in Tree Crops. *Comput. Electron. Agric.* **2012,** *83*, 11–20.

17. Chen, Y.; Marek, G. Improving SWAT Auto-irrigation Functions for Simulating Agricultural Irrigation Management using Long-term Lysimeter Field Data. *Environ. Model. Softw.* **2018,** *99*, 25–38.

18. Cheyglinted, S.; Ranamukhaarachchi, S. L.; Singh, G. Assessment of the CERES-Rice Model for Rice Production in the Central Plain of Thailand. *J. Agric. Sci.* **2002,** *137* (3), 289–298; Epub 01/23; doi: 10.1017/S0021859601001319.

19. Dar, E. A.; Brar, A. S.; Mishra, S. K.; Singh, K. B. Simulating Response of Wheat to Timing and Depth of Irrigation Water in Drip Irrigation System Using CERES-Wheat Model. *Field Crops Res.* **2017,** *214*, 149–163. doi: https://doi.org/10.1016/j.fcr.2017.09.010.

20. Dogan, E.; Clark, G. On-farm Scheduling Studies and CERES-maize Simulation of Irrigated Corn. *Appl. Eng. Agric.* **2006,** *22* (4), 509–516.

21. Enciso, J.; Unruh, B. Cotton Response to Subsurface Drip Irrigation Frequency under Deficit Irrigation. *Appl. Eng. Agric.* **2003,** *19* (5), 555–560.

22. Fang, Q.; Zhang, X.; Shao, L. Assessing the Performance of Different Irrigation Systems on Winter Wheat under Limited Water Supply. *Agric. Water Manage.* **2018,** *196*, 133–143.

23. Farahani, H. J.; Izzi, G. Parameterization and Evaluation of the AquaCrop Model for Full and Deficit Irrigated Cotton. *Agro. J.* **2009,** *101* (3), 469–476. doi: 10.2134/agronj 2008.0182s.

24. Fereres, E.; Soriano, M. A. Deficit Irrigation for Reducing Agricultural Water Use. *J. Exp. Bot.* **2006,** *58* (2), 147–159.

25. Morillo, G. Jorge; Díaz, R.; Antonio, J.; Emilio, C. Linking Water Footprint Accounting with Irrigation Management in High Value Crops. *J. Clean. Prod.* **2015,** *87*, 594–602.

26. García-Vila, M.; Fereres, E. Combining the Simulation Crop Model AquaCrop with an Economic Model for the Optimization of Irrigation Management at Farm Level. *Eur. J. Agron.* **2012,** *36* (1), 21–31. doi: https://doi.org/10.1016/j.eja.2011.08.003.

27. Gaydon, D. S.; Meinke, H.; Rodriguez, D. The Best Farm-level Irrigation Strategy Changes Seasonally with Fluctuating Water Availability. *Agric. Water Manage.* **2012,** *103*, 33–42; doi: https://doi.org/10.1016/j.agwat.2011.10.015.

28. Geerts, S.; Raes, D.; Garcia, M. Using AquaCrop to Derive Deficit Irrigation Schedules. *Agric. Water Manage.* **2010,** *98* (1), 213–216.

29. George, B.; Shende, S.; Raghuwanshi, N. Development and Testing of an Irrigation Scheduling Model. *Agric. Water Manage.* **2000,** *46* (2), 121–136.

30. Ghahraman, B.; Sepaskhah, A. R. Linear and Non-linear Optimization Models for Allocation of a Limited Water Supply. *Irrig. Drain.* **2004,** *53* (1), 39–54.

31. Goldhamer, D. A.; Fereres, E. Irrigation Scheduling Protocols Using Continuously Recorded Trunk Diameter Measurements. *Irrig. Sci.* **2001,** *20* (3), 115–125.

32. Gungula, D. T.; Kling. J. G.; Togun, A. O. CERES-maize Predictions of Maize Phenology under Nitrogen-stressed Conditions in Nigeria. *Agro. J.* **2003,** *95* (4), 892–899; doi: 10.2134/agronj2003.8920.

33. He, D.; Wang, J.; Wang, E. (Eds.). In *Modelling the Impact of Climate Variability and Irrigation on Winter Canola Yield and Yield Gap in Southwest China*, Proceedings of the 21st International Congress on Modelling and Simulation Modelling and Simulation (MODSIM) Society of Australia and New Zealand, 2015; p 314.

34. Heeren, D. M.; Werner, H. D.; Trooien, T. P. (Eds.). *Evaluation of Irrigation Strategies with the DSSAT Cropping System Model.* American Society of Agricultural and Biological Engineers (ASABE)/CSBE North Central Intersectional Meeting: St. Joseph, MI, 2006; p 214.

35. Hill, D. P. Trans-boundary Water Resources and Uneven Development: Crisis within and Beyond Contemporary India. *South Asia: J. South Asian Stud.* **2013,** *36* (2), 243–257.

36. Holzworth, D. P.; Huth, N. I.; deVoil, P. G. APSIM: Evolution Towards a New Generation of Agricultural Systems Simulation. *Environ. Model. Softw.* **2014,** *62*, 327–350; doi: https://doi.org/ 10.1016/j. envsoft.2014.07.009.

37. Hsiao, T. C.; Heng, L.; Steduto. P. AquaCrop: The FAO Crop Model to Simulate Yield Response to Water: Part III, Parameterization and Testing for Maize. *Agro. J.* **2009,** *101* (3), 448–459; doi: 10.2134/agronj2008.0218s.

38. Igbadun, H. E.; Mahoo, H. F. Irrigation Scheduling Scenarios Studies for a Maize Crop in Tanzania Using a Computer-based Simulation Model. *Agric. Eng. Int: CIGR J.* **2006,** 8; online; p 20; https://cigrjournal.org/index.php/Ejounral/article/view/679/0 (accessed Nov 30, 2019).

39. Jiang, Y.; Zhang, L.; Zhang, B. Modeling Irrigation Management for Water Conservation by DSSAT-maize Model in Arid Northwestern China. *Agric. Water Manage.* **2016**, *177*, 37–45; doi: https://doi.org/10.1016/j.agwat.2016.06.014.

40. Kharakhonova, O. *Challenges of Fresh Water Shortage. Technical Report; Красноярск, Сибирский федеральный университет*; Krasnoyarsk: Siberian Federal University, 15–25 April 2016; pages 21.

41. Krupnik, T. J.; Ahmed, Z. U. Forgoing the Fallow in Bangladesh's Stress-prone Coastal Deltaic Environments: Effect of Sowing Date, Nitrogen, and Genotype on Wheat Yield in Farmers' Fields. *Field Crops Res.* **2015**, *170*, 7–20.

42. Leite, K.; Martínez-Romero, A.; Tarjuelo, J.; Domínguez, A. Distribution of Limited Irrigation Water Based on Optimized Regulated Deficit Irrigation and Typical Metheorological Year Concepts. *Agric. Water Manage.* **2015**, *148*, 164–176.

43. Li, Q.; Qi, J.; Xing, Z. An Approach for Assessing Impact of Land Use and Biophysical Conditions Across Landscape on Recharge Rate and Nitrogen Loading of Groundwater. *Agric., Ecosyst. Environ.* **2014**, *196*, 114–124.

44. Linker, R.; Ioslovich, I.; Sylaios, G. Optimal Model-based Deficit Irrigation Scheduling Using AquaCrop: A Simulation Study with Cotton, Potato and Tomato. *Agric. Water Manage.* **2016**, *163*, 236–243; doi: https://doi.org/10.1016/j.agwat.2015.09.011.

45. Lizaso, J. I.; Boote, K. J. CSM-IXIM: A New Maize Simulation Model for DSSAT Version 4.5. *Agro. J.* **2011**, *103* (3), 766–779; doi: 10.2134/agronj2010.0423.

46. Luo, T.; Young, R.; Reig, P. *Aqueduct Projected Water Stress Country Rankings*; Technical Note; 2015; https://www.wri.org/publication/aqueduct-projected-water-stress-country-rankings (accessed Nov 30, 2019).

47. Maniruzzaman, M.; Talukder, M. S. U. Validation of the AquaCrop Model for Irrigated Rice Production under Varied Water Regimes in Bangladesh. *Agric. Water Manage.*, **2015**, *159*, 331–340; doi: https://doi.org/10.1016/j.agwat.2015.06.022.

48. Meredith, G.; Namara, R. *Impact Assessment of IWMI Contributions to Water Accounting and Water Productivity Methodologies*. Technical Report; Sri Lanka: International Water Management Institute, 2003; pages 104.

49. Moghaddasi, M.; Araghinejad, S.; Morid, S. Long-term Operation of Irrigation Dams Considering Variable Demands: Case Study of Zayandeh-rud Reservoir, Iran. *J. Irrig. Drain. Eng.* **2009**, *136* (5), 309–316.

50. Molden, D.; Murray-Rust, H.; Sakthivadivel, R.; Makin, I. *A Water-productivity Framework for Understanding and Action. Water Productivity in Agriculture: Limits and Opportunities for Improvement*; International Irrigation Management Institute (IWMI): Colombo, Sri Lanka, 2003; p 24.

51. Molden, D. *Accounting for Water Use and Productivity*; International Irrigation Management Institute, IWMI: Colombo, Sri Lanka, 1997; pages 16.

52. Montoya, F.; Camargo, D. Evaluation of AquaCrop Model for a Potato Crop under Different Irrigation Conditions. *Agric. Water Manage.* **2016**, *164*, 267–280; doi: https://doi.org/10.1016/j.agwat.2015.10.019.

53. Nikolidakis, S. A.; Kandris, D. Energy Efficient Automated Control of Irrigation in Agriculture by Using Wireless Sensor Networks. *Comput. Electron. Agric.* **2015**, *113*, 154–163.

54. Nyathi, M. K.; van Halsema, G. E. Calibration and Validation of the AquaCrop Model for Repeatedly Harvested Leafy Vegetables Grown under Different Irrigation Regimes. *Agric. Water Manage.* **2018**, *208*, 107–119; doi: https://doi.org/10.1016/j.agwat.2018.06.012.

55. Ojeda, J. J.; Caviglia, O. P. Forage Yield, Water- and Solar Radiation- Productivities of Perennial Pastures and Annual Crops Sequences in the South-eastern Pampas of Argentina. *Field Crops Res.* **2018,** *221,* 19–31; doi: https://doi.org/10.1016/j.fcr.2018.02.010.

56. Ortuño, M.; Conejero, W.; Moreno, F. Could Trunk Diameter Sensors Be Used in Woody Crops for Irrigation Scheduling? A Review of Current Knowledge and Future Perspectives. *Agric. Water Manage.* **2010,** *97* (1), 1–11.

57. Pang, X. P.; Gupta, S. C.; Moncrief, J. F. Evaluation of Nitrate Leaching Potential in Minnesota Glacial Outwash Soils Using the CERES-maize Model. *J. Environ. Qual.* **1998,** *27* (1), 75–85; doi: 10.2134/jeq1998.00472425002700010012x.

58. Peake, A. S.; Robertson, M. J.; Bidstrup, R. J. Optimising Maize Plant Population and Irrigation Strategies on the Darling Downs Using the APSIM Crop Simulation Model. *Aus. J. Exp. Agric.* **2008,** *48* (3), 313–325. doi: https://doi.org/10.1071/EA06108.

59. Perea, R. G.; Daccache, A. Modelling Impacts of Precision Irrigation on Crop Yield and In-field Water Management. *Precision Agric.* **2017,** *2017,* 1–16.

60. Perea, R.G.; García, I. F. Multiplatform Application for Precision Irrigation Scheduling in Strawberries. *Agric. Water Manage.* **2017,** *183,* 194–201.

61. Pereira, L.; Teodoro, P. Irrigation Scheduling Simulation: The Model ISAREG. In *Tools for Drought Mitigation in Mediterranean Regions.* Springer: New York, 2003; pp 161–180.

62. Popova, Z.; Kercheva, M. CERES Model Application for Increasing Preparedness to Climate Variability in Agricultural Planning—Calibration and Validation Test. *Phys. Chem. Earth, Parts A/B/C* **2005,** *30* (1–3), 125–133; doi: http://dx.doi.org/10.1016/j.pce.2004.08.026.

63. Popova, Z.; Pereira, L. S. Irrigation Scheduling for Furrow-irrigated Maize under Climate Uncertainties in the Thrace Plains, Bulgaria. *Biosyst. Eng.* **2008,** *99* (4), 587–597.

64. Pretty, J.; Bharucha, Z. P. Sustainable Intensification in Agricultural Systems. *Ann. Bot.,* **2014,** *114* (8), 1571–1596.

65. Ran, H.; Kang, S. Parameterization of the AquaCrop Model for Full and Deficit Irrigated Maize for Seed Production in Arid Northwest China. *Agric. Water Manage.* **2018,** *203,* 438–450; doi: https://doi.org/10.1016/j.agwat.2018.01.030.

66. Ran, H.; Kang, S.; Li, F. Performance of AquaCrop and SIMDualKc Models in Evapotranspiration Partitioning on Full and Deficit Irrigated Maize for Seed Production under Plastic Film-mulch in an Arid Region of China. *Agric. Syst.* **2017,** *151,* 20–32.

67. Razzaghi, F.; Zhou, Z.; Andersen, M. N.; Plauborg, F. Simulation of Potato Yield in Temperate Condition by the AquaCrop Model. *Agric. Water Manage.* **2017,** *191,* 113–123; doi: https://doi.org/10.1016/j.agwat.2017.06.008.

68. Rinaldi, M. Application of EPIC Model for Irrigation Scheduling of Sunflower in Southern Italy. *Agric. Water Manage.* **2001,** *49* (3), 185–196. doi: http://dx.doi.org/10.1016/S0378-3774 (00)00148-7.

69. Rodrigues, G. C.; Pereira, L. S. Assessing Economic Impacts of Deficit Irrigation as Related to Water Productivity and Water Costs. *Biosyst. Eng.* **2009,** *103* (4), 536–551.

70. Salemi, H.; Soom, M. A. M. Application of AquaCrop Model in Deficit Irrigation Management of Winter Wheat in Arid Region. *Afr. J. Agric. Res.* **2011,** *6* (10), 2204–2215.

71. Savenije, H. Water Scarcity Indicators; the Deception of the Numbers. *Phys. Chem. Earth, Part B: Hydrol., Oceans Atmos.* **2000,** *25* (3), 199–204.

72. Sedki, A.; Ouazar, D. Simulation-optimization Modeling for Sustainable Groundwater Development: A Moroccan Coastal Aquifer Case Study. *Water Res. Manage.* **2011,** *25* (11), 2855–2875.

73. Shang, S.; Mao, X. Application of a Simulation Based Optimization Model for Winter Wheat Irrigation Scheduling in North China. *Agric. Water Manage.* **2006,** *85* (3), 314–322.

74. Singh, A. Optimization Modeling for Conjunctive Water Use Management. *Agric. Water Manage.* **2014,** *141,* 23–29.

75. Soundharajan, B.; Sudheer, K. Deficit Irrigation Management for Rice Using Crop Growth Simulation Model in an Optimization Framework. *Paddy Water Environ.* **2009,** *7* (2), 135–149.

76. Steduto, P.; Hoogeveen, J.; Winpenny, J.; Burke, J. *Coping with Water Scarcity: An Action Framework for Agriculture and Food Security.* Food and Agriculture Organization (FAO) of the United Nations: Rome, Italy, 2017; p 83.

77. Stirzaker, R.J.; Maeko, T. C. Scheduling Irrigation from Wetting Front Depth. *Agric. Water Manage.* **2017,** *179,* 306–313.

78. Subash, N.; Shamim, M.; Singh, V. Applicability of APSIM to Capture the Effectiveness of Irrigation Management Decisions in Rice-based Cropping Sequence in the Upper-Gangetic Plains of India. *Paddy Water Environ.* **2015,** *13* (4), 325–335.

79. Sudhir, Y.; Humphreys, E. Effect of Water Management on Dry Seeded and Puddled Transplanted Rice: Part 2, Water Balance and Water Productivity. *Field Crops Res.* **2011,** *120* (1), 123–132. doi: https://doi.org/10.1016/j.fcr.2010.09.003.

80. Tan, S.; Wang, Q. Performance of AquaCrop Model for Cotton Growth Simulation under Film-mulched Drip Irrigation in Southern Xinjiang, China. *Agric. Water Manage.* **2018,** *196,* 99–113; doi: https://doi.org/10.1016/j.agwat.2017.11.001.

81. Tavakoli, A. R.; Mahdavi-Moghadam, M. Evaluation of the AquaCrop Model for Barley Production under Deficit Irrigation and Rain-fed Condition in Iran. *Agric. Water Manage.* **2015,** *161,* 136–146; doi: https://doi.org/10.1016/j.agwat.2015.07.020.

82. Thorp, K. R.; DeJonge, K. C. Methodology for the Use of DSSAT Models for Precision Agriculture Decision Support. *Comput. Electron. Agric.* **2008,** *64* (2), 276–285.

83. Toumi, J.; Er-Raki, S.; Ezzahar, J. Performance Assessment of AquaCrop Model for Estimating Evapotranspiration, Soil Water Content and Grain Yield of Winter Wheat in Tensift Al Haouz (Morocco): Application to Irrigation Management. *Agric. Water Manage.* **2016,** *163,* 219–235; doi: https://doi.org/10.1016/j.agwat.2015.09.007.

84. Wellens, J.; Raes, D. Performance Assessment of the FAO AquaCrop Model for Irrigated Cabbage on Farmer Plots in a Semi-arid Environment. *Agric. Water Manage.* **2013,** *127,* 40–47; doi: https://doi.org/10.1016/j.agwat.2013.05.012.

85. Wen, Y.; Shang, S.; Yang, J. Optimization of Irrigation Scheduling for Spring Wheat with Mulching and Limited Irrigation Water in an Arid Climate. *Agric. Water Manage.* **2017,** *192,* 33–44.

86. Whittlesey, N. Improving Irrigation Efficiency through Technology Adoption: When Will It Conserve Water? In *Developments in Water Science—50;* Elsevier: New York, 2003; pp 53–62.

87. Xiukang, W.; Zhanbin, L.; Yingying, X. Effects of Dripper Discharge and Irrigation Frequency on Growth and Yield of Maize in Loess Plateau of Northwest China. *Pak. J. Bot.* **2014,** *46* (3), 1019–1025.

88. Yeqiang, W.; Songhao, S.; Yangb, J. Optimization of Irrigation Scheduling for Spring Wheat with Mulching and Limited Irrigation Water in an Arid Climate. *Agric. Water Manage.* **2017,** *192*, 33–44.

89. Zahid, S.; Bellotti, W. (Eds.). In *Performance of APSIM-Lucerne in South Australia,* Proceedings of the 11th Australian Agronomy Conference; www.regional.org. au/au/asa/2003/c/10/bellotti.htm, 2003; p 136.

90. Zhang, D.; Li, R. Evaluation of Limited Irrigation Strategies to Improve Water Use Efficiency and Wheat Yield in the North China Plain. *PLoS One* **2018,** *13* (1), E-article I.D.: 0189989.

CHAPTER 10

SOIL MOISTURE AND NUTRIENT PATTERNS UNDER SUBSURFACE DRIP IRRIGATION FOR A SUSTAINABLE SUGARCANE INITIATIVE

M. MANIKANDAN and G. THIYAGARAJAN

ABSTRACT

Sustainable sugarcane initiative (SSI) under subsurface drip irrigation (SSDI) is gaining momentum in southern parts of India due to more output with fewer inputs. Although the benefits of SSI under SSDI are realized by farmers, yet study of water and nutrient movement in soil is lacking. To study the soil and nutrient movement for SSI under western agro-climatic zone, field experiment was conducted continuously for two different seasons (2014–2016). A variation was observed in both soil water content (SWC) and available nitrogen and potash. With an increase in horizontal spacing, SWC was decreased. However, a reverse trend was noticed for the availability of nutrients. In SSDI treatment with 100% recommended fertilizer dose and 100% pan evaporation, higher SWC and available nutrient contents were observed.

10.1 INTRODUCTION

Among various cash crops, sugarcane has attained special attraction in northern and southern regions of India. India produced 22.82% sugarcane of the world and ranked second in area (20.4%) and production (18.6%) next to Brazil. Sugarcane is grown on around 2.8% of the gross cropped area of India. The Indian national average coverage area is 5.067 million ha with a production of 36.2 million t and a productivity of 71.51 t/ha for sugarcane,[4] whereas sugarcane covers an area of about 2.631 million ha producing about 2.81 million t with an average productivity of 106.8 t/ha, in Tamil Nadu.[5]

Farmers in the western agro-climatic zone of Tamil Nadu are more interested in cultivating sugarcane crop due to high return and steady price. Although the average productivity of sugarcane in Tamil Nadu is higher than the national average, yet higher yield and maximum water and fertilizer use efficiency can be obtained only if improved packages of practices of sugarcane are adopted. To enhance the production of sugarcane with fewer inputs, sustainable sugarcane initiative (SSI) is another practical approach. SSI increases the resource use efficiencies by several folds. Under various irrigation systems, the introduction of trickle irrigation under surface has made tremendous impact both on water and fertilizer saving by 30%.[2,7,8] Proper functional design parameters of subsurface drip irrigation (SSDI) along with operation and maintenance strategies will boost up its acceptance among the farmers.[6,9,11] Hence, these aspects cannot be nullified.

The field study in this chapter was conducted to evaluate soil moisture and nutrient distribution patterns under subsurface and surface emitters for varying irrigation and fertigation levels.

10.2   METHODS AND MATERIALS

A field trail was taken up during 2014–2016 at the Agricultural Research Station—Bhavanisagar of Tamil Nadu Agricultural University (TNAU). The study was conducted to observe the varying effects of irrigation and fertilizer scheduling on soil water content (SWC) and nutrient movement. Composite soil samples at a depth of 15 cm were collected and analyzed for various physicochemical properties. The details are presented in Table 10.1. The geographical location and average meteorological parameters at the experimental site are given in Table 10.2.

Along with the basic physicochemical properties, the available values of nutrients, namely, N, P, and K were 326, 285, and 15 kg/ha, respectively. The treatment details of the experiment are given in Table 10.3.

Drip irrigation system was laid out with a lateral spacing of 1.50 m. In-line drippers of 4 L/h capacity were placed at a spacing of 60 cm for surface drip irrigation system (I_7 and I_8). Laterals with drippers of 3 L/h with 95% uniformity were spaced at 50 cm for SSDI (I_1–I_6). Common irrigations were given up to 30 days after planting. Irrigation was given once in 3 days for all treatments 30 days after planting as per treatment schedule. Sugarcane (Co.86032, variety) was used with a seed rate of 12,500 seedling/ha. Sugarcane seedlings were raised in protrays, and these were planted at 60 cm spacing.

TABLE 10.1 Soil Physicochemical Parameters.

Soil type	pH	EC (dS/m)	Infiltration rate (cm/h)	FC (%)	PWP (%)	BD (mg/m³)	OC (%)
Sandy loam	7.8	0.27	2.10	22.5	11.1	1.60	0.21

TABLE 10.2 Agro-meteorological Parameters at the Study Site.

Latitude	11.29′N
Longitude	77.08′E
Altitude, m	256 AMSL
Average rainfall, mm	716
Maximum temperature, °C	28.41
Minimum temperature, °C	18.26
Maximum relative humidity, %	84.3
Minimum relative humidity, %	42.2
Sunshine duration, h	7.4
Class A pan evaporation (PE), mm/day	5.7

TABLE 10.3 Treatment Details.

Treatment	Description of the treatment
I₁ (SSDI 100 PE,100 RDF)	SSDI + 100% pan evaporation + 100% RDF of N,K via fertigation
I₂ (SSDI 100 PE,75 RDF)	SSDI + 100% pan evaporation + 75% RDF of N,K via fertigation
I₃ (SSDI 80 PE,100 RDF)	SSDI + 80% pan evaporation + 100% RDF of N,K via fertigation
I₄ (SSDI 80 PE,75 RDF)	SSDI + 80% pan evaporation + 75% RDF of N,K via fertigation
I₅ (SSDI 60 PE,100 RDF)	SSDI + 60% pan evaporation + 100% RDF of N,K via fertigation
I₆ (SSDI 60 PE,75 RDF)	SSDI + 60% pan evaporation + 75% RDF of N,K via fertigation
I₇ (SDI 100 PE,100 RDF)	SDI + 100% pan evaporation + 100% RDF of N,K via fertigation
I₈ (SDI 100 PE,75 RDF)	SDI + 100% pan evaporation + 75% RDF of N,K via fertigation

The recommended fertilizer dose (RFD) was 275:62.5:112.5 NPK kg/ha. Entire phosphorous (P) was applied in a single dose as basal. Nitrogen (N) and potassium (K) were applied through fertigation once in 6 days. All standard package practices were followed for the growth of sugarcane. Across lateral at distances of 0, 37.5, and 75 cm and at two vertical soil depths 0–15 (d_1) and 15–30 (d_2) cm, soil samples were collected and analyzed for SWC and nutrient status.

Samples were collected at four stages for crop growth, such as germination, tillering, grand growth, and ripening. For soil moisture estimation, the samples were taken just before irrigation. SWC was computed by gravimetric method. The samples were analyzed using standard procedure for available nitrogen and potassium to know the movement of available nitrogen and potassium at different irrigation and fertigation levels.

10.3 RESULTS AND DISCUSSIONS

10.3.1 EFFECTS OF IRRIGATION ON SOIL MOISTURE DISTRIBUTION

Average SWC (in %) in relation to vertical and horizontal distances for different irrigation and fertigation schedules for SSI is presented in Table 10.4. The SWCs varied according to the treatments, where irrigations were given based on PE at different levels. SWC was decreased with increase in the horizontal distance across the lateral, and SWC was increased with increase in vertical distance irrespective of horizontal distance. SWCs were higher at 15–30 cm (d_1) depth than that at 0–15 cm (d_2) depth. SWC was found higher in the experimental plots irrigated with drip irrigation at 100% PE compared to other irrigation levels.

TABLE 10.4 Mean Soil Water Content (%) in Relation to Vertical and Horizontal Distances.

Treatment	Horizontal distance across lateral					
	0 cm		37.5 cm		75 cm	
	Vertical soil depth					
	d_1	d_2	d_1	d_2	d_1	d_2
	Mean soil water content (%)					
I_1 (SSDI 100 PE,100 RDF)	18.26	18.97	17.44	18.26	16.83	17.54
I_2 (SSDI 100 PE,75 RDF)	18.16	18.97	17.54	18.16	16.93	17.44
I_3 (SSDI 80 PE,100 RDF)	17.24	18.16	16.52	17.44	16.01	17.03
I_4 (SSDI 80 PE,75 RDF)	17.24	18.26	16.42	17.54	16.12	16.93
I_5 (SSDI 60 PE,100 RDF)	16.52	17.14	16.12	16.93	15.40	16.12
I_6 (SSDI 60 PE,75 RDF)	16.63	17.03	16.01	16.93	15.50	16.01
I_7 (SDI 100 PE,100 RDF)	17.95	19.07	17.34	18.05	16.73	17.44
I_8 (SDI 100 PE,75 RDF)	18.05	18.97	17.44	17.95	16.83	17.54

SWC was found to be higher in treatment laid with subsurface drip system than the surface drip system. Among the treatments, SSDI with 100% PE +

100% RD (I_1) recorded more soil moisture content in both cases (Fig. 10.1). The distribution of SWC under drip fertigated plots was higher near the dripper and was gradually decreased as the distance from the dripper was increased.[1,3,10] With decreasing radial distance from dripper, SWC was increased.[1]

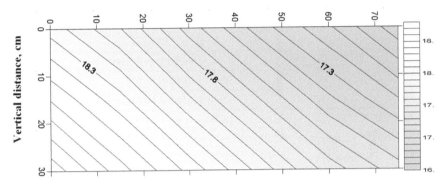

FIGURE 10.1   Soil moisture (%) pattern in relation to vertical and horizontal distances for treatments: SSDI with 100% PE + 100% RD.

10.3.2   EFFECTS OF FERTIGATION ON AVAILABLE NITROGEN AND AVAILABLE POTASSIUM IN SOIL

Available nitrogen and potassium contents (kg/ha) in relation to vertical and horizontal distances for different irrigation and fertigation schedules for SSI are presented in Tables 10.5 and 10.6.

In Tables 10.5 and 10.6, it can be observed that the available nitrogen and potassium content was increased as the distance from the laterals was increased in all treatments. The available nitrogen and potassium content was higher in the 15–30 cm (d_2) compared to 0–15 cm (d_1) soil depth. This may be due to leaching of nutrients during the movement of water to the lower depths. Among the treatments, SSDI with 100% PE + 100% RDF (I_1) recorded higher nutrient contents (Fig. 10.2).

10.4   SUMMARY

This field study revealed a decreasing trend of SWC with increase in vertical depth from the in-line source emitter, but SWC was decreased with increase in horizontal distance. For nitrogen and potassium nutrient content, the availability was increased with increase in vertical depth and was increased with increase in horizontal distance. Sugarcane performance was better with SSDI with 100% PE and 100% of RDF compared to other treatments.

TABLE 10.5 Soil Available Nitrogen (kg/ha) at Harvest in Relation to Depth and Lateral Distance.

Treatment	Horizontal distance across lateral					
	0 cm		37.5 cm		75 cm	
	Vertical soil depth					
	d_1	d_2	d_1	d_2	d_1	d_2
	Soil available nitrogen (kg/ha)					
I_1 (SSDI 100 PE,100 RDF)	337.97	342.58	350.31	354.38	364.15	367.14
I_2 (SSDI 100 PE,75 RDF)	325.21	329.83	337.56	341.63	351.40	354.38
I_3 (SSDI 80 PE,100 RDF)	317.01	321.34	328.59	332.41	341.57	344.37
I_4 (SSDI 80 PE,75 RDF)	305.05	309.38	316.63	320.45	329.61	332.41
I_5 (SSDI 60 PE,100 RDF)	327.49	331.96	339.45	343.40	352.86	355.76
I_6 (SSDI 60 PE,75 RDF)	315.13	319.60	327.10	331.04	340.50	343.40
I_7 (SDI 100 PE,100 RDF)	334.59	339.15	346.81	350.84	360.51	363.47
I_8 (SDI 100 PE,75 RDF)	321.96	326.53	334.18	338.21	347.88	350.84

Where d_1= 0–15 cm; d_2= 15–30 cm.

TABLE 10.6 Soil Available Potassium (kg/ha) at Harvest in Relation to Soil Depth and Lateral Distance.

Treatment	Horizontal distance across lateral away from dripper					
	0 cm		37.5 cm		75 cm	
	Vertical soil depth					
	d_1	d_2	d_1	d_2	d_1	d_2
	Soil available potassium (kg/ha)					
I_1 (SSDI 100 PE,100 RDF)	297.41	301.47	308.28	311.86	320.45	323.08
I_2 (SSDI 100 PE,75 RDF)	286.19	290.25	297.05	300.63	309.23	311.86
I_3 (SSDI 80 PE,100 RDF)	278.97	282.78	289.16	292.52	300.59	303.05
I_4 (SSDI 80 PE,75 RDF)	268.44	272.25	278.64	281.99	290.06	292.52
I_5 (SSDI 60 PE,100 RDF)	288.19	292.12	298.72	302.19	310.52	313.06
I_6 (SSDI 60 PE,75 RDF)	277.32	281.25	287.84	291.31	299.64	302.19
I_7 (SDI 100 PE,100 RDF)	294.44	298.46	305.19	308.74	317.25	319.85
I_8 (SDI 100 PE,75 RDF)	283.33	287.34	294.08	297.63	306.14	308.74

Where, d_1 = 0–15 cm; d_2 = 15–30 cm.

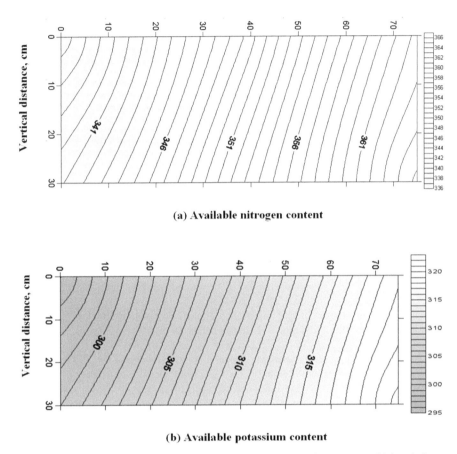

(a) Available nitrogen content

(b) Available potassium content

FIGURE 10.2   Movement of available nitrogen (a) and potassium content (b) in relation to vertical and horizontal distances for SSDI with 100% PE + 100% RD treatment.

KEYWORDS

- **drip fertigation**
- **soil moisture**
- **SSI**
- **subsurface drip irrigation**

REFERENCES

1. Ahluwalia, M. S.; Singh, B.; Gill, B. S. Drip Irrigation System: Its Hydraulic Performance and Influence on Tomato and Cauliflower Crops. *J. Water Manage.* **1993,** *1*, 6–9.
2. Camp, C. R. Subsurface Drip Irrigation: A Review. *Trans. ASABE* **1998,** *5*, 1353–1367.
3. Chakraborty, D.; Singh, A. K.; Kumar, A.; Khanna, M. Movement and Distribution of Water and Nitrogen in Soil as Influenced by Fertigation in Broccoli (*Brassica oleracea*). *J. Water Manage.* **1999,** *7*, 8–13.
4. Indiastat. Area, Production and Productivity of Sugarcane in India (2014–15); 2017a; http//indiastat.com (accessed August 15, 2019).
5. Indiastat. State Wise Area, Production and Productivity of Sugarcane in India (2014–15). 2017b; http//indiastat.com (accessed August 15, 2019).
6. Kandelous, M. M.; Šimůnek, J. Comparison of Numerical, Analytical, and Empirical Models to Estimate Wetting Patterns for Surface and Subsurface Drip Irrigation. *Irrig. Sci.* **2010,** *28*, 435–444.
7. Singh, D. K.; Rajput, T. B. S.; Singh, D. K.; Sikarwar, H. S.; Sahoo, R. N.; Ahmad, T. Simulation of Soil Wetting Pattern with Subsurface Drip Irrigation from Line Source. *Agric. Water Manage.* **2006,** *83*, 130–134.
8. Sivanappan, R. K.; Ranghaswami, M. V. Technology to Take 100 Tons Per Acre in Sugarcane. *The Kissan World* **2005,** *32*, 35–38.
9. Skaggs, T. H.; Trout, T. J.; Šimůnek, J.; Shouse. J. Comparison of HYDRUS-2D Simulations of Drip Irrigation with Experimental Observations. *J. Irrig. Drain. Eng.* **2004,** *130*, 304–310.
10. Thiyagarajan, G.; Vijayakumar, M.; Selvaraj, P. K.; Duraisamy, V. K. Performance Evaluation of Fertigation of N and K on Yield and Water Use Efficiency of Turmeric through Drip Irrigation. *Int. J. Bio Stress Manage.* **2011,** *2* (1), 69–71.
11. Thiyagarajan, G.; Vijayakumar, M.; Selvaraj, P. K.; Duraisamy, V. K. Evaluation of Irrigation Systems for Cost Reduction in Wide Spaced Sugarcane. *Int. J. Bio Stress Manage.* **2011,** *2* (4), 394–396.

PART III

Automation and Fertigation Technologies in Micro Irrigation

CHAPTER 11

AUTOMATED DRIP IRRIGATION SYSTEM FOR SWEET CORN AND CLUSTER BEAN: FIELD EVALUATION OF A LOW-COST SOIL MOISTURE SENSOR

G. RAVI BABU and N. V. GOWTHAM DEEKSHITHULU

ABSTRACT

In this chapter, drawbacks of conventional irrigation system are compared to micro irrigation. The microcontroller and wireless-based automated soil moisture sensor were developed. The chapter also evaluated the performance of microcontroller based and GSM-based automated drip irrigation system for sweet corn and cluster bean. The study also illustrated (1) moisture distribution patterns of sweet corn at different time intervals and (2) benefit cost ratio.

11.1 INTRODUCTION

There is a scarcity of water available for irrigation in India. Farmers in rural area are severally affected by this drought.[10] Only 35 million km³ (2.53% of total water on earth) fresh water is available.[12] The water level has reduced due to continuous extraction of subsurface water. In India, most farmers are using canal water with manual control.[13] To improve the crop yield and effective water utilization, several management strategies and programs have been introduced.

Micro irrigation improves crop water use efficiency (WUE) at lower cost of cultivation and reduced man power.[9] Electronic automatic systems should be introduced to assist in computing, communication, and control the irrigation requirements automatically.[7] Soil moisture sensor is the key element by which irrigation is applied in this automated irrigation system.[3,8]

Wireless sensor network (WSN) includes application, transport, network, data link, physical and power management, mobility management, and task management.[5] This system is being used in industry, health, and traffic sectors.[6] Due to reduction in the cost of sensors, researchers are also planning to introduce the WSN in agriculture sector.[11] Some wireless transceivers (such as Zigbee and Bluetooth) are more costly with high power consumption. An automatic irrigation system using Dual Tone Multiple Frequency technology was provided with temperature and moisture sensors that are connected to the microcontroller, which allows irrigation in fields requiring watering.[2]

A simple microcontroller-based system is designed and tested to control different environmental parameters. Enough field research has been carried out in the automatic irrigation system using Global System for Mobile (GSM) Communication, Bluetooth, and Zigbee technologies. However, no such product exists in market that can be used by farmers. It has been limited to its research studies only.[4]

This chapter focuses on automatic irrigation system and newly developed microcontroller moisture sensor. An efficient method of irrigation scheduling is presented in this chapter.

11.2 MATERIALS AND METHODS

The microcontroller consisted of four input/ output ports, such as P0, P1, P2, and P3 (Fig. 11.1). Each port is 8 bit wide. One of the most useful features of a microcontroller is that we can reprogram it due to flash memory. Microcontroller is the main heart of this project. The microcontroller unit (MCU) controls all the functions of other blocks of the circuit. MCU can read data from the soil moisture sensors and controls all the functions of the whole system by manipulating the data. The microcontroller detects the amount of soil moisture with four sensors.

A display unit is interfaced with the MCU for user information and displaying the condition of the field. MCU operates the motor according to the available soil moisture. The software was developed in the "C" language and was incorporated into the microcontroller by using dumper.

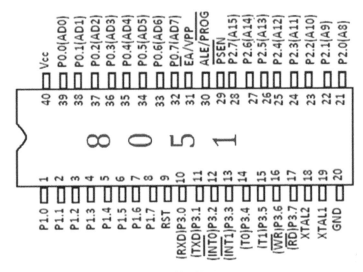

FIGURE 11.1   Pin diagram of microcontroller 8051.

11.2.1   DESIGN OF MICROCONTROLLER-BASED LOW COST SOIL MOISTURE SENSOR

To develop the microcontroller based soil moisture sensor, the components in Table 11.1 are required. These components were purchased and assembled in the box having the board as shown in Figure 11.2.

TABLE 11.1   List of Components for Developing the Microcontroller-based Soil Moisture Sensor.

Components	Specifications	Components	Specifications
Transformers: 2	Step down transformer (220/12)	Relay	220 V/3–4 A
Voltage regulator	IC 7805	Microcontroller	AT89S52
Op-amp: 4	LM741	ULN	ULN 2003
Crystal oscillator	11.0592 MHz	Water pump	
Diode	IN 4007	Switches	
LED	Red (700 nm); 470 Ω (for LED); 10K (for sensors); potentiometer (100K)	Power cables	
Resistor	1000 µf (for power supply)	Connectors	
Capacitor	33 pF (for crystal oscillator)	Solenoid valves	
LCD	16 × 2	Aluminum probes	

(a)   (b)

FIGURE 11.2 Soil moisture sensor with microcontroller and aluminum probes: (a) microcontroller-based soil moisture sensor and (b) soil moisture aluminum probes.

11.2.1.1   WORKING PRINCIPLE OF SOIL MOISTURE SENSOR

In this experiment, sensors detected the soil moisture and irrigation water was applied wherever required in the field. The developed sensor controls the irrigation water in the field using solenoid valves. The sensor present in each field stops the pump automatically through a microcontroller when the soil moisture was reached to its field capacity. Once the 70% of field capacity was attained, sensor sends the signal to the microcontroller for application of irrigation water to the required field.

11.2.1.2   WIRELESS SOIL MOISTURE SENSOR

The wireless soil moisture sensor has the property of spatially distributed autonomous to monitor physical or environmental conditions and to pass the data to a main station. The elimination of wires provides significant cost saving and creating improved reliability for future monitoring applications. The WSN (Fig. 11.3a) was initially developed in military applications and has been extended in several applications in agriculture.

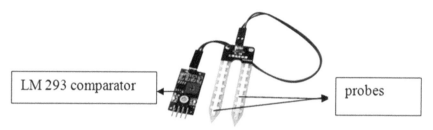

FIGURE 11.3a   Wireless soil moisture sensor.

11.2.1.3   GLOBAL SYSTEM FOR MOBILE FOR COMMUNICATION

GSM is an open, digital cellular technology used for transmitting mobile voice and data services. GSM allows network operation in roaming also so that customer can use from remote locations. This GSM facility provides key role for controlling the irrigation in the field and also sending the results to the farmer through short message service (SMS) on a mobile device, which indirectly controls the entire farm irrigation system. It is easy to implement. The system was operated according to soil moisture sensor operation.

A complete automation of an irrigation system based on SMS can be achieved through computer programming of water application (Fig. 11.3b). The SMS allowed the system to start the irrigation only when it is actually needed by the crop and is stopped when the water content has reached to preset threshold value. It was observed that selecting the right type of sensor and computer programming the automatic irrigation system will run it every day for short periods of time could save significant amount of water for the irrigation purpose.

11.2.2   FIELD EVALUATION OF AUTOMATED DRIP IRRIGATION SYSTEM

11.2.2.1   EXPERIMENTAL SITE AND CLIMATIC CONDITIONS

The experiment was conducted in 2017 at Dr. NTR College of Agricultural Engineering, Bapatla, Andhra Pradesh with latitude of 15°54.833′N and longitude of 8029.834′E for sweet corn (*Zea mays*) plots and latitude of 15°90.5915′N and longitude of 80°47.1665′E for cluster bean (*Cyamopsis tetragonoloba*) plots. The performance of sweet corn crop was evaluated under microcontroller-based soil moisture sensor and cluster bean under

GSM-based soil moisture sensor. At the site, sandy soil has more than 85% of sand and <15% of silt and clay. The soil bulk density, infiltration rate, field capacity, EC and pH were 1.49 g/cm³, 51 cm/h, 8.47%, 0.07 dSm⁻¹, and 6.45, respectively.

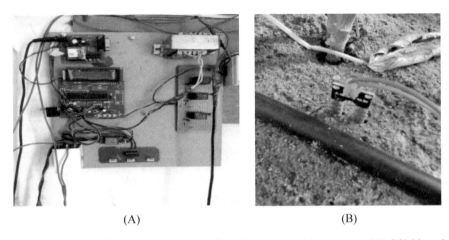

| (A) | (B) |

FIGURE 11.3b GSM-based microcontroller with a soil moisture sensor: (A) GSM based soil moisture sensor and (B) sensor placed in the field.

The maximum temperature ranged between 40 and 50°C in summer and the minimum temperature ranged between 18 and 25°C in winter. The average annual rainfall was about 115 cm. In the month of October, the Bapatla region receives a maximum rainfall of 23 cm. The average humidity and wind speed were 74.8% and 7.41 km/h, respectively.

11.2.2.2 EXPERIMENTAL LAYOUT

11.2.2.2.1 Sweet Corn Plots

Sweet corn (*Z. mays*) of variety sugar 75 was selected for this experiment. Initially 100 kg of farm yard manure (FYM) was mixed uniformly in the field. The sweet corn seeds were sown on January 25, 2015 (Figs. 11.4 and 11.5).

The plot was thoroughly wetted for 2 days before sowing. The sweet corn seeds were sown in the experimental field with recommended seed rate of 10 kg/ha in these plots. The entire field with an area of 1330 m² was divided into three following plots, each having dimensions of 12 m × 35 m (Fig. 11.4):

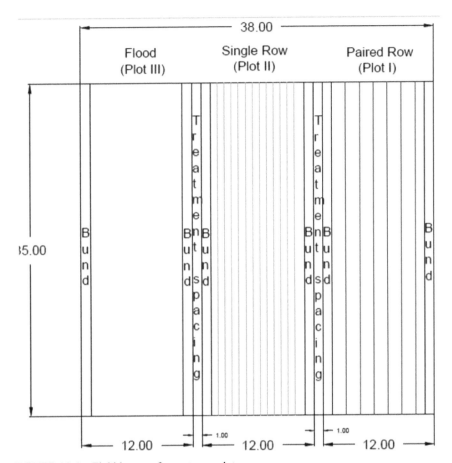

FIGURE 11.4   Field layout of sweet corn plot.

NOTE: All dimensions are in mm.

LEGEND: Green (0.75 m lateral spacing); pink (1.5 m lateral spacing).

Plot I (the paired row, drip irrigation): The plant to plant spacing was 0.2 m, paired row spacing of 0.4×1.1 m². The drip lateral was placed with 1.5 m spacing and the distance between dripper to plant was maintained 0.2 m on both sides of the lateral and recommended dose of fertilizer was applied manually.

- *Plot II* (single row: drip irrigation): The spacing of 0.2×0.75 m²; recommended dose of fertilizer was applied manually.
- *Plot III* (Flood irrigation): The spacing of 0.2×0.75 m²; recommended dose of fertilizer applied manually.

(a)   (b)

FIGURE 11.5 Laying of laterals for sweet corn: (a) laying of lateral for every row and (b) laying of lateral for paired row.

The drip irrigation system was arranged as shown in Figure 11.6.

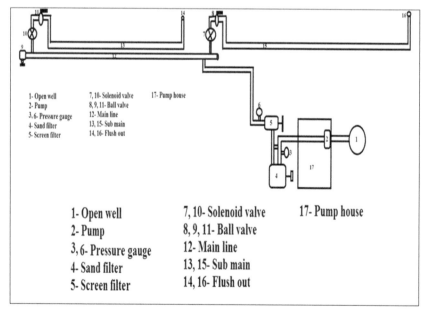

1- Open well	7, 10- Solenoid valve	17- Pump house
2- Pump	8, 9, 11- Ball valve	
3, 6- Pressure gauge	12- Main line	
4- Sand filter	13, 15- Sub main	
5- Screen filter	14, 16- Flush out	

FIGURE 11.6 Drip irrigation system with accessories at the experimental site.

11.2.2.2.2   Cluster Bean Plots

Cluster bean (*C. tetragonoloba*) of variety Guar was selected for this experiment. Initially, 30 kg of FYM was mixed uniformly in the field. The cluster bean seeds were sown on March 24, 2016. The plot was thoroughly wetted for 2 days before sowing. The entire field with an area of 406 sq. m^2 was divided into three following plots (Fig. 11.7):

- *Plot I* (flood irrigation): Area of 7 × 16 m^2 and plant spacing of 0.4 × 0.4 m^2.
- *Plot II* (drip irrigation): The paired row with an area of 4.4 × 16 m^2. The plant-to-plant spacing was 0.4 m with paired row spacing of 0.3 × 0.5 m^2. The distance between dripper to plant was 0.15 m on both sides of lateral and spacing of lateral to lateral was 0.8 m.
- *Plot III* (drip irrigation): The third single row with area of 5 × 22 m^2 and plant spacing was 0.4 × 0.4 m^2.

11.2.2.3   CULTIVATION PRACTICES USED FOR SWEET CORN AND CLUSTER BEAN CROPS

- *Fertilizer application:* Based on standard recommended dosage of fertilizer application for sweet corn, 58.5 kg of phosphorus fertilizer (single super phosphate) was applied as a basal dose. The nitrogen (61 kg of urea) and potassium (14.2 kg of Muriate of potash) fertilizers were applied in three splits manually at the required stages of crop growth of sweet corn crop. Similarly for cluster bean, 10 kg of phosphorus fertilizer (single super phosphate) was applied as a basal dose. The nitrogen (20 kg of urea) and potassium (4.2 kg of Muriate of potash) fertilizers were applied in three splits manually at 30, 45, and 60 days of sowing, respectively.
- *Micronutrients for sweet corn:* Growth was applied for the crop (2 mL/L) with a power sprayer.
- *Insecticides and pesticides used for sweet corn:* Carbofuran (1 kg) and monocrotophos (2 mL/L) were applied for sweet corn crop for preventing stem-bore disease.

11.2.2.4   DETAILS OF OBSERVATIONS FOR SWEET CORN AND CLUSTER BEAN CROP

Biometric parameters (such as plant height, branches per plant, number of leaves, and root growth), yield, cob characteristics, and wetting patterns of drip-irrigated treatments were evaluated.

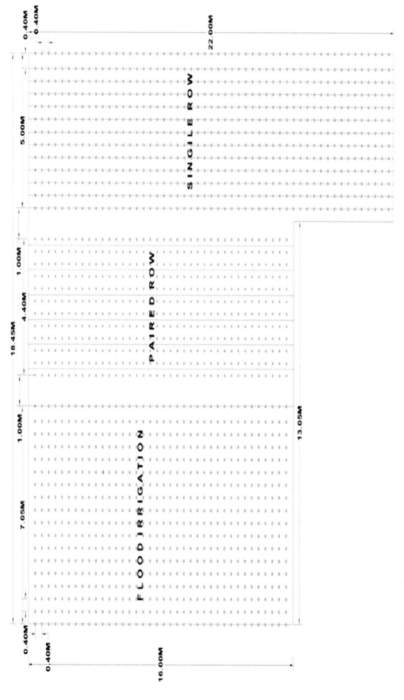

FIGURE 11.7   Field layout of cluster bean plots.

11.2.2.5   SWEET CORN AND CLUSTER BEAN: ESTIMATION OF CROP WATER REQUIREMENTS AND IRRIGATION SCHEDULING

For sweet corn and cluster bean, the amount of water applied and irrigation scheduling were calculated using FAO CROPWAT software[1] using Penman–Monteith method. For the crop data, the K_c values were taken from the FAO 56, and for the soil data, soil texture was sandy soil. The climatic data for the past 10 years were used to calculate potential evapotranspiration.

11.2.2.6   EMISSION UNIFORMITY

The following Karmeli and Keller equation was used to estimate the emission uniformity (EU) for drip irrigation system:

$$EU = 100 \times \left[1.0 - \left\{ \frac{(1.27\ C_v)}{n^{0.5}} \right\} \right] \times \left[\frac{q_m}{q_a} \right] \qquad (11.1)$$

where q_m and q_a are minimum and average discharge of drippers and C_v is the coefficient of variation.

11.2.3   MOISTURE DISTRIBUTION PATTERNS

A study was carried out on the moisture distribution patterns under drip irrigation. Trench method was used to determine moisture distribution for every 30 min up to 3 h. The horizontal and vertical spread of water was recorded (Fig. 11.8) at different time intervals. Moisture distribution patterns were drawn with a Surfer software using the recorded observations (Fig. 11.9).

11.2.4   WATER USE EFFICIENCY

For calculating WUE, the yield obtained and the amount of water applied for each plot were recorded.

$$WUE = \frac{\text{yield}}{\text{amount of water applied}} \qquad (11.2)$$

FIGURE 11.8   Measuring the water distribution under drip irrigation.

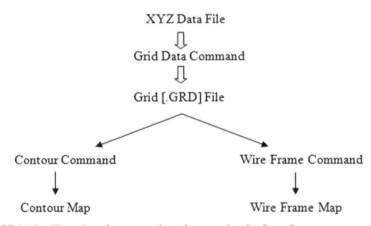

FIGURE 11.9   Flow chart for preparation of maps using Surfer software.

11.2.5   *COST ECONOMICS OF WIRELESS SOIL MOISTURE SENSOR IN DRIP IRRIGATION*

The objective of economic analysis was to compare the use of various inputs of production and income. The study included the cost of all electronic components (Fig. 11.10) and devices of the developed soil moisture sensor

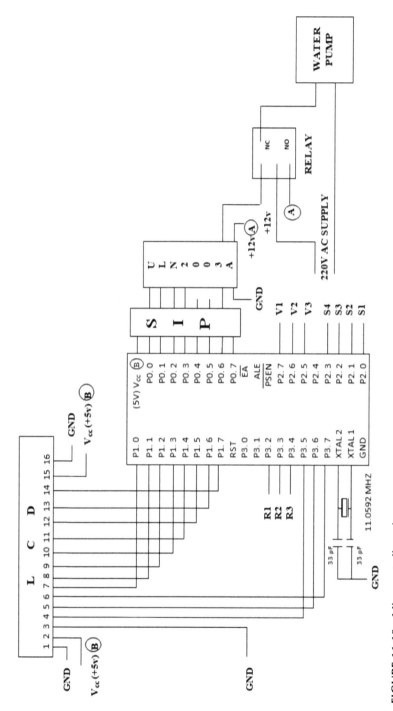

FIGURE 11.10   Microcontroller unit.

integrated with drip head work via GSM and SMS to provide information to farmer of real-time status of agricultural field. The expenditure incurred from field preparation to harvest was worked out and expressed in Rs./ha.

The yield of cluster bean was computed as kg/ha, and the total income was worked out based on the prevailing minimum market rate of Rs. 15.00/kg. The benefit cost ratio (BCR) was determined using the following equation:

$$BCR = \frac{\text{Gross return}(Rs./ha)}{\text{Total cost of cultivation}(Rs./ha)} \qquad (11.3)$$

11.3   RESULTS AND DISCUSSIONS

11.3.1   DEVELOPMENT OF SOFTWARE FOR THE MICROCONTROLLER

The high signal (logic 1) appears on the output pin of the sensor of a particular field entertained by the microcontroller when the crop needs water (i.e., field capacity of soil reaches to 70%). By knowing the position of the pin on which signal appears, the microcontroller switch ON the RELAY (i.e., water pump) connected at PORT 0. Now water applied in the particular field by opening of solenoid valve. After completion of watering (i.e., field soil moisture content reaches to the field capacity), the sensor sends low signal (logic 0) to microcontroller and switches OFF the water pump. Now microcontroller starts sensing the signal at another PORT. In developing the software in the microcontroller, program was planned for maximum of four sensors.

11.3.2   DEVELOPMENT OF LOW COST MICROCONTROLLER — BASED SOIL MOISTURE SENSOR

In the development of low-cost microcontroller-based soil moisture sensor, 5 V DC power supply is needed for all electronic operations. Step down transformer, rectifier, voltage regulator, and filter circuit are required for generation of 5 V DC power. The total cost of the microcontroller-based soil moisture sensor was Rs. 3755 compared to the Rs. 100,000 for commercial available automatic unit for the irrigation system. The microcontroller based soil moisture sensor works on the circuit diagrams as shown in Figures 11.11–11.14.

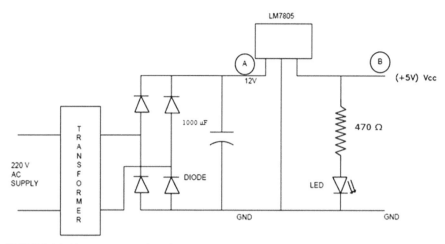

FIGURE 11.11 Power supply unit.

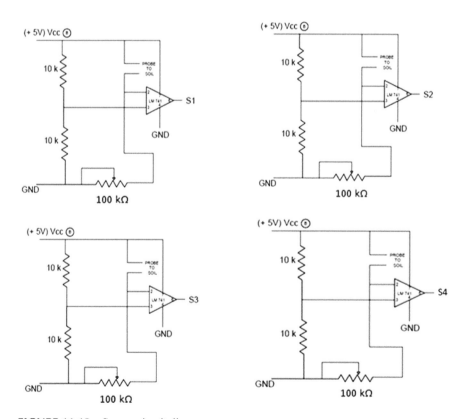

FIGURE 11.12 Sensor circuit diagrams.

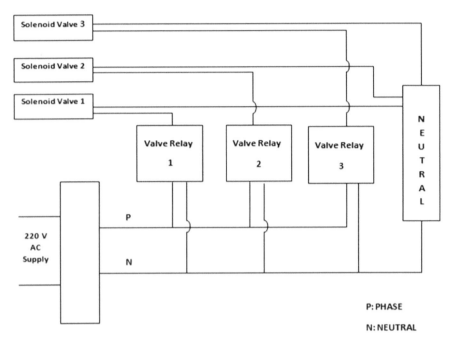

FIGURE 11.13   Power supply to solenoid valves.

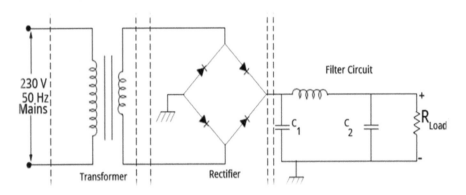

FIGURE 11.14   Power supply unit.

11.3.3   DEVELOPMENT OF LOW-COST MICROCONTROLLER GSM-BASED WIRELESS MOISTURE SENSOR

In the development of low-cost microcontroller GSM-based wireless soil moisture sensor, 5 V DC power supply is needed for all electronic operations. The power supply works on the circuit diagrams as shown in Figures 11.15–11.17.

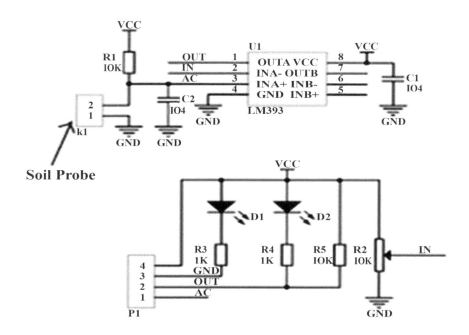

FIGURE 11.15 Sensor circuit diagram.

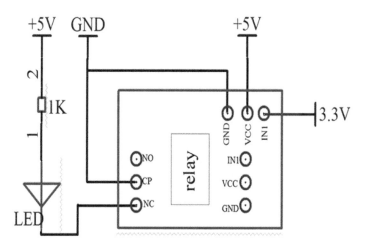

FIGURE 11.16 Relay circuit diagram.

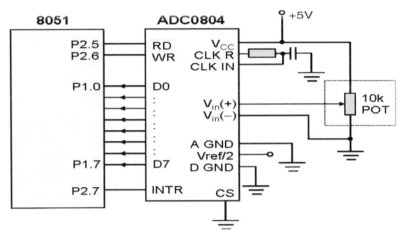

FIGURE 11.17   ADC interfacing with microcontroller circuit diagram.

11.3.4   FIELD EVALUATION OF AUTOMATED DRIP IRRIGATION SYSTEM

11.3.4.1   CROP WATER REQUIREMENT FOR SWEET CORN

The total water requirement of sweet corn crop based on 10 years' climate data was 466.4 mm from sowing of the seed (January 25th) to harvest of the crop (April 24th). The rainfall was not considered in irrigation calculations (effective rainfall = 0) as there was no rainfall during the growing period (Fig. 11.18).

| ETo station | Bapatla | | | | | Crop | sweet corn |
| Rain station | Bapatla | | | | | Planting date | 25/01 |

Month	Decade	Stage	Kc	ETc	ETc	Eff rain	Irr. Req.
			coeff	mm/day	mm/dec	mm/dec	mm/dec
Jan	3	Init	1.15	4.79	33.5	0.0	33.5
Feb	1	Init	1.15	5.04	50.4	0.0	50.4
Feb	2	Deve	1.15	5.27	52.7	0.0	52.7
Feb	3	Deve	1.13	5.56	44.4	0.0	44.4
Mar	1	Mid	1.12	5.84	58.4	0.0	58.4
Mar	2	Mid	1.12	6.18	61.8	0.0	61.8
Mar	3	Mid	1.12	6.51	71.6	0.0	71.6
Apr	1	Late	1.09	6.71	67.1	0.0	67.1
Apr	2	Late	1.03	6.60	26.4	0.0	26.4
					466.4	0.0	466.4

FIGURE 11.18   Crop water requirements for sweet corn.

The total water requirement was also calculated for the present climate data from January to April 2015. It shows a total water requirement of 450.7 mm from sowing of the seed (January 25th) to harvest of the crop (April 24th). There is only a variation of 15.7 mm of total water requirement from previous 10 years' average data to the present 4 months' data. Actual water applied in all irrigation treatments for sweet corn crop was given in Table 11.2. Water saving of 36% was observed in automated drip irrigation systems for sweet corn crop.

TABLE 11.2   Actual Water Applied for Sweet Corn under Different Irrigation Systems.

Treatment	Water applied during sowing (mm)	Water applied during crop growth (mm)	Total water applied (mm)	Percentage of water saving
Flood	50	470	520	–
Single row drip	50	282	332	36
Paired row drip	50	282	332	36

11.3.4.2   CROP WATER REQUIREMENT FOR CLUSTER BEAN CROP

The total water requirement of cluster bean crop based on 10 years' climate data is 456.7 mm from sowing of the seed (March 24, 2017) to harvest of the crop (June 21, 2017). The rainfall was considered in irrigation calculations (effective rainfall = 5.3 mm) as there was rainfall during the growing period (Fig. 11.19). Actual water applied in all irrigation treatments for cluster bean crop is given in Table 11.3. Water saving of about 19% was observed in the automated drip irrigation systems for cluster bean.

Crop Water Requirements							
ETo station bapatla						**Crop**	cluster bean
Rain station bapatla						**Planting date**	24/03
Month	Decade	Stage	Kc	ETc	ETc	Eff rain	Irr. Req.
			coeff	mm/day	mm/dec	mm/dec	mm/dec
Mar	3	Init	0.50	2.65	21.2	0.1	21.1
Apr	1	Init	0.50	2.74	27.4	0.2	27.2
Apr	2	Deve	0.57	3.24	32.4	0.2	32.2
Apr	3	Deve	0.77	4.48	44.8	0.4	44.4
May	1	Deve	0.97	5.79	57.9	0.5	57.4
May	2	Mid	1.10	6.72	67.2	0.6	66.6
May	3	Mid	1.10	6.77	74.5	0.8	73.6
Jun	1	Mid	1.10	6.80	68.0	1.1	67.0
Jun	2	Late	1.01	6.29	62.9	1.3	61.6
Jun	3	Late	0.90	5.53	5.5	0.1	5.5
					461.8	5.3	456.7

FIGURE 19   Crop water requirements for cluster bean.

TABLE 11.3   Actual Water Applied for Cluster Bean under Different Irrigation Systems.

Treatment	Water applied during sowing (mm)	Water applied during crop growth (mm)	Total water Applied (mm)	Percentage of water saving
Flood	50	457	507	–
Single row drip	50	360	410	19.13
Paired row drip	50	360	410	19.13

11.3.4.3   EMISSION UNIFORMITY: SWEET CORN AND CLUSTER BEAN FIELDS

The EU was calculated by measuring the discharge for selected drippers and using Karmeli and Keller equations. The all calculations are given in Table 11.4. As the length of sweet corn plot is more, the EU is less.

TABLE 11.4   Emission Uniformity: Sweet Corn and Cluster Bean.

Crop	Standard deviation, S	Coefficient of variation, C_v	Emission uniformity (%)
Sweet corn	5.80	0.028	92.87
Cluster bean	1.87	0.014	98.80

11.3.4.4   BIOMETRIC PARAMETERS FOR SWEET CORN AND CLUSTER BEAN CROP

Biometric parameters for two crops under different treatments are presented in Tables 11.5 and 11.6.

TABLE 11.5   Biometric Parameters for Sweet Corn under Different Treatments.

Parameter	Flood irrigation	Single row drip	Paired row drip
Plant height (m)	2.22	2.51	2.04
Root depth (cm)	24	28	22

TABLE 11.6   Biometric Parameters for Cluster Bean under Different Treatments.

Parameter	Flood irrigation	Single row drip	Paired row drip
Plant height (m)	0.6	0.9	0.75
Root depth (cm)	20	25	23
No. of beans/plant	45	60	50

11.3.4.4.1   Cob Characteristics

To study the sweet corn cob characteristics, 1 m^2 area was selected in each treatment. The observed cob characteristics were number of kernel rows per cob, number of kernels per cob, cob length, cob diameter, and the fresh cob weight. The number of cobs in 1 m^2 area was collected, and the average of those cobs is presented in Table 11.7.

TABLE 11.7   Cob Characteristics for Different Irrigation Treatments.

Treatment	No. of kernel rows/cob	No. of kernels/cob	Cob diameter	Cob length	Individual fresh cob weight
	–	–	cm		g
Flood irrigation	16	656.8	5.15	20.26	367.6
Single row drip	16.8	755.2	5.51	20.54	405.2
Paired row drip	14.8	558.8	3.824	18.8	226.6

11.3.4.4.2   Sweet Corn Yield under Different Irrigation Treatments

The total yield of sweet corn yield for different experimental plots was calculated and is presented in Table 11.8. The yield from the flood irrigation plot, single row drip plot, paired row drip plot was 7.43, 7.93, and 6.48 t/ha, respectively. The corn yield in single row drip plot was higher compared to the yield in other experimental plots. The increased yield may be obtained by the efficient application of water at right time near the root zone by low-cost microcontroller-based soil moisture sensor, which is installed in the field and supplies water automatically whenever there is need of water to the plant, thus providing favorable conditions for plant growth.

TABLE 11.8   Sweet Corn Yield under Different Irrigation Systems.

Irrigation system	Plot size (m^2)	Yield per plot (kg)	Yield (kg/ha)	Yield (t/ha)
Flood	12× 35	312	7429	7.43
Single row drip	12× 35	333	7929	7.93
Paired row drip	12× 35	272	6476	6.48

1 t = 1000 kg.

11.3.4.4.3   Yield Response of Cluster Bean under Different Irrigation Treatments

The cluster bean yield for different experimental plots was calculated and is presented in Table 11.9. The yield from the flood irrigation plot, paired row drip plot, and single row drip plot was 4.4, 6.0, and 7.5 t/ha, respectively. The increased yields could be obtained in a single row drip due to efficient application of water at right time by low-cost GSM-based wireless soil moisture sensor, which is present in the field and supplies water automatically whenever there is need of water to the plant that creates favorable conditions for plant growth. The percent increase in yield was 70.84 and 36.00% in single row drip and paired row drip system, respectively, compared to flood method.

TABLE 11.9   Cluster Bean Yield under Different Irrigation Systems.

Irrigation system	Plot size (m²)	Yield per plot (kg)	Yield (kg/ha)	Yield (t/ha)	Percentage increase in Yield
Flood	7×16	49.5	4,419	4.41	–
Paired row drip	4.40×16	42.0	5,966	5.97	36.00
Single row drip	5×22	82.5	7,500	7.50	70.84

11.3.5   MOISTURE DISTRIBUTION PATTERNS IN DIFFERENT TREATMENTS FOR SWEET CORN AND CLUSTER BEAN

11.3.5.1   WETTING SPREAD

The horizontal wetted perimeter and depth of penetration of water were measured for every 30 min in all three treatments during the water application. The water was applied according to soil moisture sensor in drip systems. With increase in time of operation, the wetted diameter and penetration depth were increased in relative to time (Fig. 11.20).

11.3.5.2   MOISTURE DISTRIBUTION GRAPHS FOR SWEET CORN

In a light sandy soil, the ponding is very small and was hardly observed. The shape of the soil moisture distribution graph was close to conical for all the irrigation scheduling (i.e., for the time of 30, 60, and 90 min [Figure 11.21]).

The soil moisture distribution pattern showed highest water content near the emitter in all scheduling techniques.

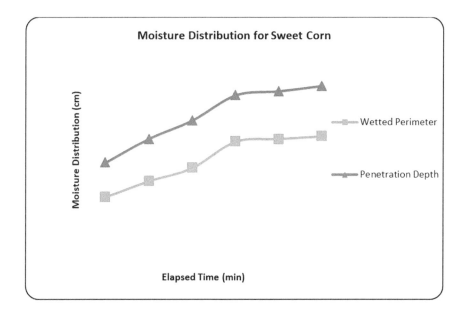

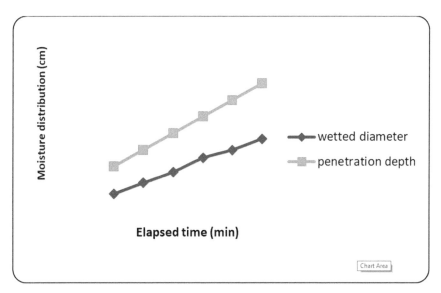

FIGURE 11.20   Relationship between wetted perimeter and penetration depth with elapsed time for sweet corn and cluster bean.

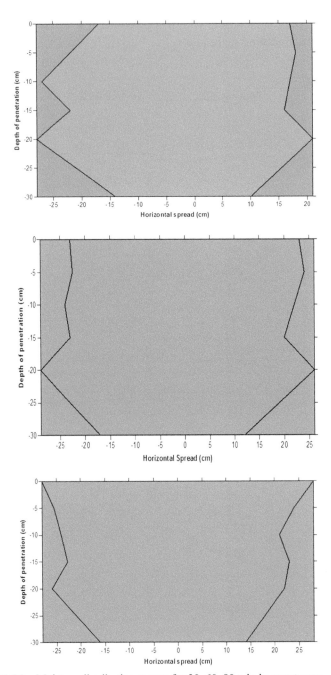

FIGURE 11.21   Moisture distribution curves for 30, 60, 90 min in sweet corn.

11.3.5.3   SOIL MOISTURE DISTRIBUTION PATTERNS FOR SWEET CORN

In order to study the moisture distribution pattern around the plant, the soil moisture content was measured at different depths below the soil surfaces at varying radial distances from the plant. Soil moisture distribution patterns at the 30, 60, and 90 min during irrigation are shown in Figures 11.22–11.24. At 30 min during irrigation, amount of moisture content was decreased as the distance from the plant was increased due to lateral spacing and the moisture content near the plant was 8.3%.

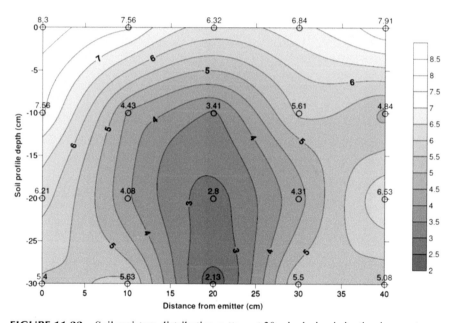

FIGURE 11.22   Soil moisture distribution pattern at 30 min during irrigation in sweet corn.

The moisture content at 10 cm depth near the plant was 7.56%. The moisture content was reduced from 7.56 to 5.4% at a depth of 10–30 cm near the plant. The percentage decrease in moisture content near the plant was 34.9%. At a distance of 10 cm from the plant, the moisture content was increased from 4.43 to 5.63% at a depth of 10–30 cm from surface. The percentage decrease in moisture content at a distance of 10, 20, and 30 cm from the plant was 25.5, 66.2, and 19.5%.

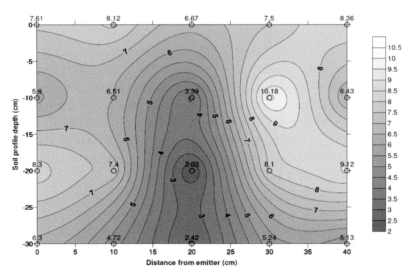

FIGURE 11.23 Soil moisture distribution pattern at 60 min during irrigation in sweet corn.

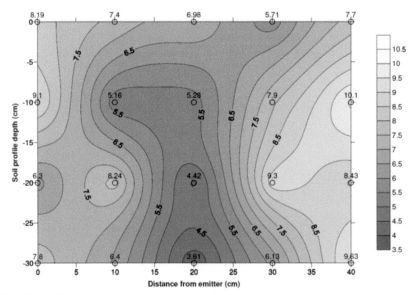

FIGURE 11.24 Soil moisture distribution pattern for 90 min during irrigation in sweet corn.

At 90 min during irrigation, the moisture content near the plant is 8.19% and reduction in moisture content near the plant is 4.76%. At a distance of 30 and 40 cm from the plant, the moisture content was increased from 5.71 to 6.13% and 7.7 to 9.63% at a depth of 30 cm from the surface. The percentage decrease in moisture content at a distance of 10 and 20 cm from the plant was 13.5 and 48.2%.

11.3.6   WATER USE EFFICIENCY FOR SWEET CORN

Table 11.10 indicates that highest WUE for sweet corn was 23.88 kg/ha mm in single row drip method followed by 19.51 in paired row drip and 14.29 in flood method, respectively. In the paired row drip system, higher WUE was observed than the flood. There was 36% water saving even though yield is less compared to the flood method.

TABLE 11.10   WUE for Sweet Corn under Different Irrigation Methods.

Treatment	Yield (kg/ha)	Water applied (mm)	WUE (kg/ha mm)
Flood irrigation	7429	520	14.29
Single row drip	7929	332	23.88
Paired row drip	6476	332	19.51

11.3.7   WATER USE EFFICIENCY FOR CLUSTER BEAN

Table 11.11 indicates that highest WUE for cluster bean was 18.29 kg/ha mm in a single row drip method followed by 14.55 in paired row drip and 8.71 in flood method. In a single row drip system, higher WUE was observed than flood with 19.13% water saving.

TABLE 11.11   WUE of the cluster bean under different irrigation methods.

Treatments	Yield (kg/ha)	Water applied (mm)	WUE (kg/ha mm)
Flood irrigation	4419	507	8.71
Paired row drip	5966	410	14.55
Single row drip	7500	410	18.29

11.3.8   COST ECONOMICS OF GSM-BASED WIRELESS SOIL MOISTURE SENSOR FOR AUTOMATION OF DRIP IRRIGATION SYSTEM

The cost of the GSM-based wireless soil moisture sensor was Rs. 3424.00 for all components, such as microcontroller, soil moisture sensors, GSM supporter, and transformer (Table 11.12). This cost is lower compared to the automatic system supplied by the drip irrigation companies costing Rs. 100,000. The BCR of cluster bean crop is shown in Table 11.13.

TABLE 11.12   List and Cost of Components for the GSM-Based Sensor.

Component	Quantity	Unit cost (Rs.)	Total cost (Rs.)
Microcontroller	1	64	64
24 V relays	4	20	80
Soil moisture sensors	4	150	600
LCD display	1	140	140
Analog-to-digital converter (ADC)	1	120	120
GSM supporter board	1	900	900
Transformer 24 V	1	70	70
Connectors	2	125	250
Two-wired coils	2	600	1200
Total			3424.00

TABLE 11.13   Benefit Cost Ratio for Cluster Bean Crop in All Three Treatments.

Parameter	Single row	Paired row	Flood
Annual cost of drip irrigation system (depreciation, interest, repair, and maintenance), Rs./ha, A	33,000	22,000	0
Cost of cultivation, Rs./ha, B	20,000	20,000	32,000
Total cost (A + B), Rs/ha, E	53,000	42,000	32,000
Crop yield, t/ha, C	7.5	6.0	4.4
Selling price, Rs./t @ Rs. 15/kg D	15,000	15,000	15,000
Income from produce (C × D), Rs., F	112,500	90,000	66,000
Benefit cost ratio, F/E	2.12	2.14	2.06

11.4   SUMMARY

For crop duration of 90 days for sweet corn and cluster bean, the water applied was calculated by FAO CROPWAT method in flood irrigation method, whereas water application was based on the operation of microcontroller-based automatic soil moisture sensor for both single-row and paired-row drip irrigation methods for sweet corn and cluster bean. The following conclusions are drawn from this study:

- A low-cost microcontroller-based automatic soil moisture sensor device was developed based on the basic principle of electrical conductivity. Soil moisture sensor was operated according to the software included in the microcontroller basing on the soil field capacity.
- The EU of drip irrigation was 92.87% in sweet corn.
- The best yield of sweet corn was 7.93 t/ha in soil moisture sensor-based irrigation with each row lateral spacing compared to paired row drip method. In this case, efficient use of water was due to plant rows nearer to the drip lateral.
- Highest WUE in sweet corn was 23.88 kg/ha mm in single row drip method followed by 19.51 for paired row drip and 14.29 for flood method, respectively. In paired row drip system, there was 36% water saving compared to flood method.
- The amount of water applied in flood irrigation system for cluster bean crop was 455 mm based on FAO CROPWAT method in addition to 50 mm water for sowing of cluster bean. Hence, in total, 507 mm of water was used in flood method.
- A percentage of 19.13 of water saving is observed in drip method for cluster bean crop.
- The EU of drip irrigation was 98.8% in the case of cluster bean.
- The plant length and root depth were 90 m and 25 cm, respectively, in a single row drip for cluster bean compared to other irrigation methods.
- The highest yield of cluster bean was 7.5 t/ha in a single row drip method because of efficient use of water by the plant compared with flood method and paired row drip method.
- As increase in time of operation wetting diameter and penetration depth increased in relation with time, moisture content is found to be more after irrigation than before irrigation.
- Highest WUE in cluster bean crop was 18.29 kg/ha-mm in a single row drip method compared to 14.55 kg/ha-mm for paired row drip and 8.71 kg/ha-mm in flood irrigation, respectively. In a single row drip system, WUE caused 19.13% water saving.
- The total annual cost for the drip system was Rs. 20,000 for cluster bean cultivation. The BCR of 2.14 was observed for paired row followed by 2.12 for single row and the lowest BCR was 2.06 in flood irrigation.
- The several sections in the field can be operated in an automated drip irrigation system using wireless soil moisture sensor with the help of solenoid valves.

- The GSM-based wireless soil moisture sensor used in automation drip irrigation system is designed to operate the pump automatically based on soil field capacity.

KEYWORDS

- **benefit cost ratio**
- **drought**
- **microcontroller**
- **relay**
- **surfer**
- **water use efficiency**

REFERENCES

1. Allen, R. G.; Pereira, L. S.; Raes, D.; Smith, M. *Crop Evapotranspiration: Guidelines for Computing Crop Water Requirements*. FAO Irrigation and Drainage Paper 56, 1998; pp 65–182.
2. Amit, V.; Ankit, K.; Avdesh, S.; Atul, S. Automatic Irrigation System Using DTMF. *Int. J. Eng. Tech. Res.* **2014,** *2014,* 157–159.
3. Anand, K. S. Vyavasaya Panchamgam. *ANGRAU* **2014,** *2014,* 38–45.
4. Atodaria, V. H.; Tailor, A. M.; Shah, Z. N. SMS Controlled Irrigation System with Moisture Sensors. *Ind. J. Appl. Res.* **2013,** *3* (5), 506–507.
5. Chavan, C. H.; Karande, P. V. Wireless Monitoring of Soil Moisture, Temperature & Humidity using Zigbee in Agriculture. *Int. J. Eng. Trends Technol.* **2014,** *11* (10), 493–497.
6. Chung, W. Y.; Jocelyn, F. V.; Janine, T. Wireless Sensor Network Based Soil Moisture Monitoring System Design. *Position Papers 2013 Federated Conf. Comput. Sci. Info. Syst.* **2013,** *2013,* 79–82.
7. Divya, V.; Umamakeswari, A. An Intelligent Irrigation System with Voice Commands and Remote Monitoring of Field. *J. Artif. Int.* **2013,** *6* (1), 101–106.
8. Hani, A. M.; Hany, M. M.; El-Hagarey, M. E.; Ahmehd, S. H. Using Automation Controller System and Simulation Program for Testing Closed Circuits of Mini-sprinkler Irrigation System. *Open J. Model. Simul.* **2013,** *1,* 14–23.
9. Patil, K. M.; Bhaskar, P. C. Microcontroller Based Adaptive Irrigation System Using WSN for Variety Crops and Development of Insect Avoidance System for Better Yield. *Int. J. Res. Eng. Technol.* **2014,** *3* (7), 308–312.
10. Prasad, K. S. S.; Nitesh, K.; Sinha, N. K.; Kumar, P. S. Water-saving Irrigation System Based on Automatic Control by Using GSM Technology. *Middle-East J. Sci. Res.* **2012,** *12* (12), 1824–1827.

11. Rahim, K.; Ali, I.; Asif, S. M.; Mushtaq, A.; Zakarya, Md. Wireless Sensor Network Based Irrigation Management System for Container Grown Crops in Pakistan. *World Appl. Sci. J.* **2013,** *24* (8), 1111–1118.

12. Reddy, K. Y.; Gorantiwar, S. D. Performance Evaluation of Solar Photovoltaic Pumping System. *Andhra Agric. J.* **1997,** *44* (1 & 2), 1–5.

13. Sakthipriya, N. An Effective Method for Crop Monitoring Using Wireless Sensor Network. *Middle-East J. Sci. Res.* **2014,** *20* (9), 1127–1132.

14. Suresh, R.; Gopinath, S.; Govindaraju, K.; Devika, T.; Vanitha, N. S. GSM Based Automated Irrigation Control Using Raingun Irrigation System. *Int. J. Adv. Res. Comput. Commun. Eng.* **2014,** 3 (2), 5654–5657.

CHAPTER 12

FERTIGATION TECHNOLOGY FOR HORTICULTURAL AND FIELD CROPS

RAJA GOPALA REDDY, KAMLESH N. TIWARI, and SANTOSH D. T.

ABSTRACT

Application of solid fertilizers leads to loss of nutrients resulting in low yield compared to the liquid fertilizers that can save time, labor, fertilizer, etc. This chapter discusses knowledge of indigenous materials for multiple uses (such as methods of application, compatibility, and calculation of fertilizer need, and system design and its application). More detailed research is needed to know the exact amount of fertilizer requirement for individual crop in different soil types and climatic conditions and to improve production, quality, and yield of horticultural crops. Alternate modern techniques are required for precise application with low-cost and long-lasting methods and instruments for better yield with minimum loss of fertilizer.

12.1 INTRODUCTION

The commercial fertigation system started in the middle of the 20th century.[13] The main factor behind promoting a fertigation system is a drip irrigation system, which was started in Israel and spread over the world.[11,17] Recently, a fertigation has become a valuable tool in major horticultural crops. It has gained popularity because of its efficient use in fertilizer management, time-saving, labor management, and potentially better control over crop performance. In fertigation, soluble fertilizers are applied near the root zone along with water, which leads to better use efficiency of water and nutrients to obtain the maximum yield.

Application of solid fertilizers leads to loss of nutrients in the form of volatilization, leaching into deeper layers promoting nutrient deficiency, low yield, and contamination of natural resources. However, frequent application of water-soluble fertilizers near the crop root zone uniformly results in rapid uptake of nutrients by plants and overcomes many issues related to nutrient uptake, time, labor, etc.

This chapter discusses various aspects of fertigation in horticultural and orchard crops along with suitable examples.

12.2 FERTIGATION

Fertigation is the application of fertilizers along with irrigation water through the irrigation system. Fertilizers are evenly distributed in the irrigation water and, therefore, nutrients are available to plants with improved nutrient use efficiency. In the fertigation system, efficiency of fertilizer has increased to 80–90% because they are readily soluble and immediately available to plants without any loss. The various advantages and limitations of the fertigation are as follows:

- Fertilizers are applied evenly to all plants by fertigation near the crop root zone so that it provides more absorption and better crop growth.
- It increases the yield by about 25–50% compared with the conventional and other methods of fertilization.
- It saves labor, water, and nutrients. Fertigation increases fertilizer use efficiency (FUE) by 80–90%.
- In fertigation, time and energy use is also reduced significantly.
- To avoid leaching of nutrients due to excessive water in the root zone, (1) the fertigation is allowed in the middle of irrigation; (2) water irrigated should be free from sand and other clogging agents, which cause clogging of drippers leading to uneven nutrient distribution near the root zone; (3) use only water-soluble fertilizers or liquid fertilizers— avoid using phosphorus, calcium, magnesium, bicarbonates fertilizers because most of these are insoluble in water and lead to clogging of drippers; (4) use only corrosion-resistant fertigation equipments; and (5) avoid potential chemical backflow into the water supply source.

12.3 TYPES OF FERTILIZERS

12.3.1 NITROGEN

Research studies have reported that water-soluble nutrients through drip irrigation can increase crop yield by 10% compared to basal application of solid fertilizers.[7]

Plants absorb nitrogen in the form of nitrates (NO_3^-) and ammonia (NH_4^+). These ions are released only after decomposition and mineralization of organic matter. It shows better growth and development of plant, increases protein content in grains and quality of fodder. Nitrogen deficiency leads to stunted growth, and pale yellow color develops near the tip of the leaf blade later developing into V-shaped pattern on margins and then whole leaf is turned to a yellow color. Excess application of nitrogen shows dark green leaves that cause delaying in maturity and effects on the product quality. It also causes more incidence of pest attack and susceptible to diseases and lodging. N fertilizers suitable for fertigation are urea, ammonium sulfate, calcium nitrate (CAN), and ammonium nitrate. Urea is the most suitable and easily water-soluble compared with other sources of solid fertilizers.

12.3.2 PHOSPHORUS

Phosphorous (P) is the second most important macronutrient absorbed by plants largely in the form of $H_2PO_4^-$ from the soil solution. It plays an important role in the structural component of cell, chloroplast, and mitochondria. P is involved in energy transfer and transformation of sugars, starch, and other nutrients to various parts of the plant. It plays an important role in cell division, the growth of meristem, root, seed, fruit development and helps in N fixation in nodule bacteria of legume crops. Normally, P deficiency is seen in soils having low organic matter, acidic, and calcareous nature. Deficiency symptoms observed on older leaves with a dark blue-green color start from the tip toward the base later turning into reddish-purple color.

12.3.3 POTASSIUM (K)

Potassium chloride (KCl) is the cheapest source of K. Distribution of potassium in the soil is uniform in the case of drip fertigation.[6] It is easily soluble in water and does not cause any salt precipitation. Mostly, KCl is commonly used in fertigation, which plays an important role in regulating the stomatal function, increases drought tolerance, neutralizes organic anions, enhances crop quality and shelf life of fruits, reduces lodging of crops, and imparts disease resistance. Deficiency of K appears on older leaves with leaf burnt along margins, which gradually progress inward showing burning appearance. K deficiency reduces the crop yield without appearance of definite symptoms and this phenomenon is called hidden hunger. For fertigation, it is available in the form of potassium nitrate, KCl, and monopotassium phosphate.

12.3.4 CALCIUM

Calcium (Ca) is an important secondary macronutrient. It is available in different sources, such as calcium carbonate, calcium sulfate, and CAN. However, for fertigation, CAN is the only source of Ca, which can be applied through fertigation to regulate Ca-related problems in fruits and vegetables. It plays an important role in cell division, cell elongation and it moves in the plant only in one direction toward apical part.[16] In acid soil to control this, lime is added to soil to maintain the soil pH and to promote root proliferation.

12.3.5 MAGNESIUM (Mg)

Magnesium (Mg) plays an important role in chlorophyll formation, which imparts green color to leaves and also helps in protein synthesis and activation of ATP compounds. In clayey soils, the percentage availability of Mg is about 6% due to the presence of montmorillonite clay particles, which supply magnesium to the soil solution very slowly. Magnesium is available in different sources, such as magnesium sulfate, dolomite, and magnesium phosphate; however, water-soluble magnesium fertilizers are magnesium nitrate and magnesium sulfate.

12.3.6 SULFUR (S)

Sulfur (S) is an essential nutrient for plant growth for the production of S containing amino acids like cysteine, cystine, methionine and for protein synthesis. It is a constituent of certain vitamins like thiamine, biotin, acetyl coenzyme-A, and ferredoxin. Sulfur improves pungency in onion, garlic and increases oil content in oilseed crops. It meets the requirement of plants in the form of potassium sulfate, ammonium sulfate, and magnesium sulfate.

12.3.7 IRON

Iron (Fe) is an important micronutrient for respiratory and photosynthetic reaction. Iron activates some enzymes, including nitrogenase, aminolevulinic acid synthase, and coproporphyrinogen oxidase. It is immobile in the plant, which leads to iron deficiency. Normally, iron deficiency is seen in calcareous soils with high pH and soils with low organic matter content.

12.3.8   MANGANESE

Plants absorb manganese (Mn) as Mn^{2+} that helps in the electron transfer in photosystem-II of photosynthesis. It influences auxin levels and maintains chloroplast membrane structure. It protects plant cells against ill effects of superoxide free radicles. It normally shows deficiency symptoms on plants grown in calcareous soils and soil with low organic matter content. Manganese deficiency symptoms show interveinal chlorosis on middle and lower leaves due to translocation to younger leaves. It shows grayish-yellow/brown spots of minute size in interveinal areas on basal parts of the leaf and later moves to tip.

12.3.9   ZINC (Zn)

Zinc (Zn) is one of the micronutrients that is essential for agricultural and horticultural crops. Most of the Indian soils are Zn deficient. Normally, plant contains 20–150 mg of Zn/kg and the critical level of Zn deficiency is between 15 and 20 mg Zn/kg depending on crop type and variety. It is not translocated from the older to younger leaves quickly. Hence, Zn-deficiency symptoms can be observed initially on younger second/third fully matured leaves. The affected leaves are white to pale yellow developed between veins and margins at the base of leaf and later extend toward the tip. In the case of severe deficiency, leaf tissue is developed into brown necrosis and leaves become very brittle and give a rosette appearance.

12.3.10   COPPER (Cu)

Copper (Cu) deficiency is seen in crops growing in peat soils. Plants contain 5–20 mg/kg, which is immobile in plants. Its deficiency symptoms show initially on the young mature leaves. It causes senescence, male sterility, delay in flowering and loss of grain yield.

12.3.11   BORON (B)

Boron (B) plays an important role in the transport of sugars, synthesis of the cell wall, lignification, metabolism of phenols, IAA, RNA, transport of sugars, pollen germination, and growth of pollen tube. It regulates the opening of stomata and imparts drought resistance to crops. Boron is immobile in plants; hence its absence is seen in younger leaves. Under severe conditions, leaves become thick and brittle, internodes become short, and shoot tips die causing the growth of lateral shoots.

12.3.12   MOLYBDENUM

It shows deficiency symptoms mainly in acid soils on old or middle leaves as an interveinal chlorotic mottling followed by marginal necrosis. Under severe conditions, midribs are converted into whip/tail like commonly known as whiptail seen in cauliflower and broccoli.

12.4   SOLUBILITYAND COMPATIBILITY OF FERTILIZERS AND RECOMMENDATIONS IN ACIDIC AND ALKALINE SOILS

12.4.1   SOLUBILITY

Solubility is the quantity of nutrient, which is soluble in a unit amount of water. Normally, nitrogen and potash fertilizers do not have solubility problem. However, phosphate fertilizers such as DAP and SSP do not readily dissolve in water.[8,11,22] Variations of temperature significantly affect the solubility of fertilizers. The solubility decreases with the decrease in temperature. The FUE for different irrigation systems[3,5] is given in Table 12.1. Tables 12.2–12.7 provide the solubility limit (g/L) of nitrogenous, potash and phosphate, and micronutrient fertilizers.

TABLE 12.1   Fertilizer Use Efficiency (FUE) under Irrigation Systems.

Fertilizer	Fertilizer use efficiency (%)		
	Conventional	Drip	Fertigationa
Nitrogen	30–50	65	95
Phosphorus	20	30	45
Potash	50	60	80

[a]*95% nutrients applied through fertigation are taken up by plants.*

TABLE 12.2 Solubility of Nitrogenous Fertilizers.

Types of fertilizer	Solubility (g/L)	N content (%)
Ammonium nitrate	1920	34
Ammonium sulfate	750	21
Calcium nitrate	1290	15.5
Urea	1100	46

TABLE 12.3 Suitability of Nitrogenous Fertilizers for Fertigation.

Fertilizer	Analysis (N:P:K)	pH (g/L at 20°C)
Urea	46:0:0	5.8
Potassium nitrate	13:0:46	7.0
Ammonium sulfate	21:0:0	5.5
Urea ammonium nitrate	32:0:0	−
Ammonium nitrate	34:0:0	5.7
Mono ammonium phosphate	12:61:0	4.9
Calcium nitrate	15:0:0	5.8
Magnesium nitrate	11:0:0	5.4

TABLE 12.4 Solubility of Potash Fertilizers.

Fertilizer	K content (%)	Solubility (g/L)
K_2SO_4	50	110
KCl	60	340
KNO_3	44	133

TABLE 12.5 Potash Fertilizers Suitable for Fertigation.

Fertilizer	Analysis (N–P–K)	pH (g/L at 20°C)	Other nutrients
Potassium chloride[a]	0–0–60	7.0	46% Cl
Potassium nitrate	13–0–46	7.0	13% N
Potassium sulfate[b]	0–0–50	3.7	18% S
Potassium thiosulfate[c]	0–0–25	−	17% S
Monopotassium phosphate	0–52–34	5.5	52% P_2O_5

[a]*Only white.*

[b]*Only Fertigation grade.*

[c]*Liquid.*

TABLE 12.6 Phosphorus Fertilizers for Fertigation.

Fertilizers	Analysis (N:P:K)	pH (g/L at 20°C)
H_3PO_4	0–52–0	2.6
KH_2PO_4	0–52–34	5.5
$NH_4H_2PO_4$	12–61–0	4.9

TABLE 12.7 Solubility of Micronutrients.

Fertilizer	Content (%)	Solubility (g/L)
Solubor	20 B	220
$CuSO_4$	25 Cu	320
$FeSO_4$	20 Fe	160
$MgSO_4$	10 Mg	710
$(NH_4)_2MoO_4$	54 Mo	430
$ZnSO_4$	36 Zn	965
$MnSO_4$	27 Mn	1050

12.4.2 SOLUBILITY OF PHOSPHATIC FERTILIZERS

- Calcium phosphate (P + Ca): insoluble
- Magnesium ammonium phosphate (P + ammonium + magnesium): insoluble.
- Iron phosphate (P + iron): insoluble.
- If (Ca + Mg) in water is >50 ppm and bicarbonate >150 ppm, then even good sources of P like polyphosphates get precipitated. If bicarbonate is <100 ppm, the, (Ca + Mg) can go up to 75 ppm.

12.4.3 COMPATIBILITY

Combining two or more fertilizers sometimes may lead to the formation of precipitates.[9,10,21] Hence, they can be dissolved in two separate tanks to prepare solution. Figure 12.1 gives a compatibility chart of different water-soluble fertilizers.

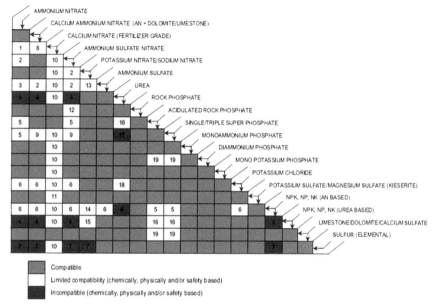

FIGURE 12.1 Fertilizer compatibility chart; Note: To avoid clogging in drip line due to insoluble precipates, Calcium fertilizers should not be mixed with fertilizers containing sulfates or phosphates in the same concentrated solution.

Source: http://customhydronutrients.com/fertilizer-concentrates-compatibility-ezp-6.html

12.4.3.1 INTERACTION AMONG THE COMPATIBILITY OF NUTRIENTS IN FERTILIZERS

The solubility product differs with different materials and this must be considered while preparing the fertilizer solution.[2,9] Table 12.8 indicates fertilizer evaluation for suitability to fertigation. Recommended fertilizers for fertigation in acidic (4.5–6.5) and neutral—alkaline soils (6.5–8.5) are shown in Table 12.9.

TABLE 12.8 Fertilizer Evaluations for Suitability to Fertigation.

Property	NH_4NO_3	$(NH_4)_2SO_4$	K_2SO_4	KCl	KNO_3	H_3PO_4	MAP
Compatibility	Good	Poor	Poor	Medium	Medium	Medium	Good
Corrosion	Medium	Poor	Poor	Poor	Good	Poor	Medium
Precipitation	Low	High	High	Low	Low	Low	High
Solubility	High	Medium	Low	Medium	Medium	High	Medium

TABLE 12.9   Recommended Fertilizers for Fertigation in Acidic (4.5–6.5) and Neutral—Alkaline Soils (6.5–8.5).

Nutrient	Neutral—basic soils	Acidic—neutral soils
	(pH 6.5–8.5)	(pH 4.5–6.5)
Nitrogen	–	Ammonium nitrate (NH_4NO_3)
		Potassium nitrate (KNO_3)
		Calcium nitrate ($Ca(NO_3)_2$)
		Urea
	Ammonium sulfate ($(NH_4)_2SO_4$)	–
	Ammonium phosphate ($NH_4H_2PO_4$)	
Phosphorus	Phosphoric acid (H_3PO_4)	Mono-ammonium phosphate (KH_2PO_4)
	–	Ammonium polyphosphate
Potassium	–	Muriate of potash (KCL)
		Potassium sulfate (K_2SO_4)
		Potassium nitrate (KNO_3)
Secondary nutrients	–	Calcium nitrate $Ca(NO_3)_2$
		Magnesium nitrate $Mg(NO_3)_2$
		Potassium sulfate (K_2SO_4)
Micronutrients	–	B as a boric acid
		Mo as sodium molybdate
		EDTA complex with Cu, Zn, Mo, Mn
	Fe–EDDHA	Fe–EDTA
	Fe–DTPA	–

The solubility of the mixture fertilizer depends on the formation of precipitates as follows:

- $CaSO_4$ (gypsum) will precipitate, when CAN reacts with any sulfate:

$$Ca(NO_3)_2 + (NH_4)_2SO_4 \lozenge CaSO_4\downarrow + \ldots\ldots\ldots$$

- Calcium phosphate will precipitate, when CAN reacts with any phosphate:

$$Ca(NO_3)_2 + NH_4PO_4 \lozenge CaHPO_4 \downarrow + \ldots\ldots\ldots$$

- Magnesium phosphate will precipitate, when magnesium reacts with mono-ammonium phosphate:

$$Mg(NO_3)_2 + NH_4H_2PO_4 \lozenge MgHPO_4\downarrow + \ldots\ldots\ldots$$

- Ammonium sulfate with KCl or KNO_3 will cause the formation of K_2SO_4 precipitates:

$$SO_4(NH_4)_2 + KCl \text{ or } KNO_3 \lozenge K_2SO_4 \downarrow + \text{………}$$

- Phosphorus reacts with iron resulting in the precipitation of iron phosphates:

12.5 FERTIGATION METHODS

12.5.1 PRESSURE DIFFERENTIAL METHOD

In the fertilizer tank (Fig. 12.2), water flow is decreased in the control head due to the pressure differential and a fraction of water is diverted through a fertilizer solution. To divert a sufficient amount of solution through a micro-tube of 9–12 mm in diameter, a gradient of 0.1–0.2 bar is required. The fertilizer tank is made of galvanized iron/stainless steel and is coated with polyester inside and outside to make it corrosion resistant. Solid fertilizers are dissolved in the tank very slowly and mixed with the flowing water. The concentration of the solution is mostly constant till the fertilizer is leftover in the tank. This is a simple and economical method without the use of external energy (pump). The disadvantage of this unit is the nutrient concentration of fertilizer solution and fertilizer injection rate cannot be precisely regulated. Refill the fertilizer tank before each application and it is not suitable for automation.

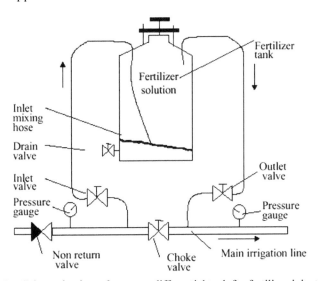

FIGURE 12.2 Schematic view of pressure differential tank for fertilizer injection.

12.5.2   VENTURI INJECTION METHOD

In Venturi system (Fig. 12.3), water flows through a constricted tube due to which the water velocity is increased and a negative pressure is developed at this point, which allows sucking of the fertilizer solution from the open-source of the tank with a tube connected to the constricted section. The device is normally made up of plastic materials that are free from corrosion problems. Venturi injection rate depends on pressure loss that varies from 10 to 75% of the initial pressure depending on the injector type and operating conditions. To minimize the pressure loss, Venturi device requires excess pressure for a constant flow and uniform nutrient application over time to maintain pressure and booster pump can be attached. Suction rate depends on inlet pressure loss. The inlet pressure is about 33%, but in case of double stage requires only 10% pressure loss. The suction rate varies from 100 to 2000 mL/h. Injectors are attached to inline or on the bypass system. This entire process of injection of fertilizer solution is based on Bernoulli's principle (eq 12.1), assuming inlet and outlet ports of Venturi are at the same elevation, and total energy at a point is constant.

$$\frac{P}{\gamma} + \frac{V^2}{2g} + Z = \text{constant} \qquad (12.1)$$

where P is the pressure, V is the flow velocity, Z is the elevation head, γ is the unit weight of water, and g is the acceleration due to gravity = 9.81 m/s².
    Considering inlet and outlet ends of a Venturi injector are designated as points 1 and 2, respectively, and they are located at same elevations, therefore $Z_1 = Z_2$. Equation 12.1 is reduced to the following equation:

$$\frac{P_1}{\gamma} + \frac{V_1^2}{2g} = \frac{P_2}{\gamma} + \frac{V_2^2}{2g} \qquad (12.2)$$

or

$$\frac{1}{\gamma}\left(P_1 - P_2\right) = \frac{1}{2g}\left(V_2^2 - V_1^2\right) \qquad (12.3)$$

The change in velocity V_2 from V_1 due to constriction in Venturi pipe diameter causes the suction (change in pressure from P_1 to P_2) below the atmospheric pressure, thereby causing suction of fertilizer solution. Figure 12.3 explains the principle of operation of a Venturi injector for application of fertilizer in a banana field (Fig. 12.4).

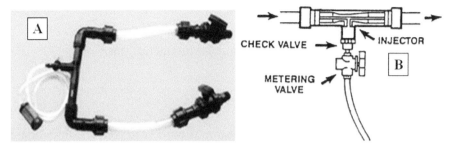

FIGURE 12.3   Venturi injectors ((a) normal view and (b) schematic view).

FIGURE 12.4   Layout of a banana plantation with fertigation unit.

Advantages

1. Easy installation of the device, low cost, and uniform mobility of nutrient solution.
2. It is simple in operation with a wide range of suction rates and it does not require external energy.
3. It is compatible with an automated irrigation system.
4. The injection can be controlled with a metering valve.
5. Suitable for both proportional and quantitative fertilization.

Disadvantages

1. Chances for a significant pressure loss.
2. Injection rates are affected by pressure fluctuations.

12.5.3 FERTILIZER INJECTION PUMP

In this method, a fertilizer injection pump is used to lift the fertilizer liquid from fertilizer tank, and it is injected into the pipe that carries irrigation water under desired pressure for proper mixing of the solution with water (Fig. 12.5). Fertilizer pump is driven by electricity, internal combustion (IC) engines, tractor power take off (PTO). Among these, hydraulic pumps are multipurpose with low maintenance and operational cost. The movement of piston injects the fertilizer solution into the irrigation system. Positive injection pumps include diaphragm pumps, piston pumps with single or multiple types, roller, and gear pumps. Multiple pump units can be used, where multiple types of fertilizers are required and to reduce precipitation problems. Adjustment of the length of stroke of piston pump or selection of appropriate diameter of pulley can regulate injection of pumps to achieve the preferred rate of application. The system should be flushed with clean water at least once after injecting the fertilizer solution. The major advantages of the system are flexibility and high discharge rate, the system does not add to the head loss in a pressurized irrigation system, and it maintains a constant concentration of nutrients throughout fertilizer application. The disadvantages are the high cost of the equipment and high operation and maintenance costs.[16]

12.5.4 BACKFLOW PREVENTION

It is a device in the form of check—valves to protect the irrigation source from contamination by chemicals. It is the main fertigation safety device, which is used to prevent backflow (Fig. 12.6).

The backflow is caused due to the higher downstream pressure and lower upstream pressure in the fertigation system. Higher downstream pressure caused by pumps and lower upstream is because of water usage exceeds the amount of water supplied. Such situations can be avoided by installing double check valves or special valves that combine check valves with reduced pressure zones commonly known as reduced pressure principle backflow prevention valves.[4,8,17] Back siphoning is caused by a negative pressure created by vacuum in the fertigation system. It is similar to drinking water with a straw. Such situations can be avoided by installing check valves, vacuum relief valves, or vacuum breaker valves.

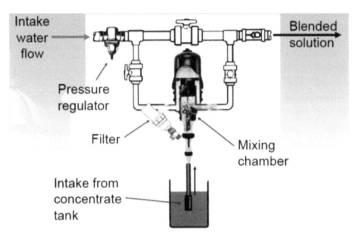

FIGURE 12.5   A positive displacement pump for fertigation.

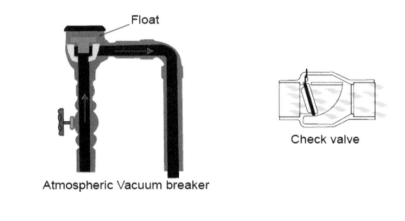

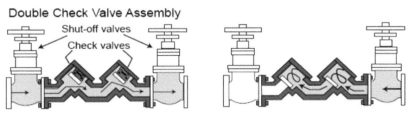

FIGURE 12.6   Backflow prevention devices like vacuum breakers and double check valves.

12.5.5   SAFETY PRECAUTIONS FOR SUCCESSFUL FERTIGATION SYSTEM

1. Variations in operating pressure should be minimized to achieve thorough mixing of fertilizer solution and irrigation water.
2. Before using mixed fertilizers, check the compatibility chart to avoid corrosion and precipitation that may clog the emitters.
3. Strainers should be used to avoid the entry of solid/undissolved particles into the system.
4. Change the water pH based on the pH of fertilizer solution.
5. Other chemicals like pesticides, herbicides, and chlorine are not injected in combination with fertilizer solution.
6. Irrigation and fertilizer injection pumps should be compatible to prevent entry of fertilizer in the irrigation line. The control for the motors of both the pumps should be electronically interlocked.
7. Check valves should be installed to prevent the backflow. Double check valves and vacuum relief valves should be installed to prevent from draining of admixture of fertilizer and water or tapping back into the water source.

12.6   FERTIGATION SYSTEM DESIGN

For drip irrigation system, wetted soil portion is partial and is unsaturated; the root distribution system is restricted and concentrated. The nutrients are washed out from the crop root zone very quickly, therefore uninterrupted.

Factors crucial for effective fertigation design are (1) estimation of available nutrients in soil, (2) estimation of amount of fertilizer required, (3) frequency of fertigation, (4) fertilizer tank capacity, (5) irrigation water requirement, (6) drip system capacity, (7) fertigation duration, (8) estimation of concentration of nutrients in irrigation water, and (9) injection rate.

12.6.1   FERTIGATION PARAMETERS

For the effective fertigation, four following criteria must be considered:

- the kind of fertilizer required by the crop,
- climatic condition,
- physiochemical properties of soil, and
- the exchange capacity of the soil.

With fertigation system, the plant nutrients are applied along with irrigation water at a given interval at the required concentration. With this approach, leaching of nutrients especially nitrogen is decreased and it improves the FUE. Fertigation improves the soil conditions by lowering the variation of soil salinity, which can be favorable for salt-sensitive crops.[19,20] In general with fertigation, chemical fertilizers can be dissolved and applied for a long time on a sustainable basis without contaminating soil and water.

12.6.1.1 ESTIMATION OF FERTILIZER REQUIREMENT

The requirement of fertilizers changes according to the stage of plant growth. The quantity of nutrients applied during each fertigation and during the peak crop production stage depends on the fertigation intervals, soil type, crop nutrient demand, and its availability in the soil. The required amount of fertilizer may be estimated by the following equation:

$$F_n = (R - A_n) \times F_{cf} \tag{12.4}$$

where F_n is the nutrient requirement, kg/ha; R is the recommended dose of fertilizer for the crop, kg/ha; A_n is the available fertilizer in the soil, kg/ha; and F_{cf} is the fertilizer correction factor.

To determine the quantity of fertilizer to be injected into the system for each setting, the area irrigated in each setting of the lateral line is obtained by multiplying the length of the lateral coverage and the move of the lateral. The quantity of fertilizer to be injected is calculated as follows:

$$D_f = \frac{D_s \times D_e \times N_s \times W_f}{10,000} \tag{12.5}$$

where D_f is the amount of fertilizer per setting, kg; D_s is the distance between sprinklers, m; D_e is the distance between laterals, m; N_s is the number of sprinklers; and W_f is the recommended fertilizer dose, kg/ha.

Numerical example: A lateral has 20 sprinklers spaced 10 m apart. The laterals are spaced 20 m on the main line. Determine the amount of fertilizer to be applied at each setting when the recommended fertilizer dose is 100 kg/ha.

Solution: Using eq. 12.5, $D_s = 10$ m, $D_e = 20$ m, $N_s = 20$, and $W_f = 100$ kg/ha. The quantity of fertilizer to be applied is estimated as follows:

$$D_f = \frac{10 \times 20 \times 20 \times 100}{10,000} = 40 \text{ kg}$$

12.6.1.2   FREQUENCY OF FERTIGATION, DURATION, AND CAPACITY OF FERTILIZER TANK

12.6.1.2.1   Fertigation Frequency

Fertigation frequency may be varied from crop to crop and time of application. It may be daily, weekly, or monthly depending on the age of the plants. Mostly, frequency depends on soil type, irrigation scheduling, system design, crop nutrient requirement, and the farmer's preference. The application of water should be monitored, as fertilizer application is linked to water application.[12] In many cases, it is extremely important that either during the irrigation or subsequent irrigation, the fertigation is not subjected to leaching.

12.6.1.2.2   Injection Duration Fertigation

The duration of maximum injection depends on the soil type and crop requirements of water and nutrients. The maximum period of 45–60 min[1] is normally recommended with adequate time for flushing of fertilizer residues from the laterals before finishing pump operation. There should be a sufficient duration for uniform distribution of nutrients throughout the fertigated soil surface. Better to inject for a long duration to leave enough time to flush chemicals out of the system rather than in a "slug," where concentrated fertilizer solutions are injected in less than 45 min.

Injection duration must be complete within allowable time period to prevent the supply of excess water, because it may leach plant nutrients beyond the root zone. Also, excess water may damage the roots and plants by saturating the soil.

12.6.1.2.3   Fertilizer Concentration

The actual nutrients concentration needed in the irrigation water depends on the crop type. Many systems have flowing water along with maintaining desired concentration of a chemical in the system. It requires injecting a fertilizer solution with appropriate rate to maintain the desired concentration level. The fertilizer concentration in the irrigation water is estimated as follows[6]:

$$C_f = \frac{100\,F_r}{T_r \times I_d} \tag{12.6}$$

where C_f is the concentration of nutrients in irrigation water, ppm; T_r is the ratio between fertilization time and irrigation time; I_d is the gross irrigation depth, mm; and F_r is the rate of fertilizing irrigation cycle, kg/ha.

12.6.1.2.4 Fertilizer Injection Rate

The injection of fertilizer into the system is based on the concentration of the liquid nutrient and the required amount of fertilizer to be applied during irrigation. It may be scheduled by the size of the injector or the irrigation system flow rate. Injection time should be limited to prevent overapplication of water that will leach chemicals. To calculate the desired injection rate, the following equation can be used if all the parameters are known:

$$q_{fi} = \frac{F_r \times A}{c \times t_r \times I_t} \qquad (12.7)$$

where q_{fi} is the injection rate of liquid fertilizer solution into the system, L/h; F_r is the rate of fertilizing (quantity of nutrient to be applied) per irrigation cycle, kg/ha; A is the area irrigated (ha) in time, I_t; c is the concentration of nutrient in liquid fertilizer, kg/L; t_r is the ratio between fertilization time and irrigation time; and I_t is the duration of irrigation, h.

12.6.1.2.5 Fertilizer Tank Capacity

The fertilizer tank capacity depends on the number of fertilizer combinations to be added, which in turn will depend on either the total amount of fertilizer to be applied or on the length of the fertigation period. Low-cost tanks are practical, where an injection pump or Venturi used. A large tank helps to store fertilizer tank in which storage ranges from 30 to 600 L. This is not enough, because some fertilizers need larger tank with capacity of 300–600 L. The stock solution is prepared based on the solubility of the fertilizers used. Normally, the highest concentration is not desirable, and it is recommended to prepare stock solution with slightly lesser concentration. The fertilizer tank capacity is computed by using the following equation:

$$C_t = \frac{F_r \times A}{c} \qquad (12.8)$$

where C_t is the capacity of fertilizer tank, L; F_r is the rate of fertilizing per irrigation cycle, kg/ha; A is the area irrigated (ha) in time, I_t; and c is the concentration of nutrient in liquid fertilizer, kg/L.

12.7   WATER QUALITY IN FERTIGATION

Water plays an important role in fertigation. Impure water contains iron, calcium, magnesium that can clog the emitters to reduce the application rate and water distribution. Chemical and biological materials also clog the emitters, which are the main source of obstruction.

12.7.1   WATER CONTAINING IRON

Irrigation water contains a soluble iron, which appears as a gelatinous film of brown-reddish slime precipitate in the form deposits on the dripping material, thus resulting in clogging of the emitters. It occurs in places, where groundwater pH is less than 7.0 and there is a lack of dissolved oxygen in the water. Fe^{2+} is a water-soluble and main source for slime formation. Bacteria like Enterobacter, Pseudomonas, and Sphaerotihus react with Fe^{2+} through an oxidation process and form Fe^{3+} as an insoluble compound and create sticky gelatinous film, which clog the emitters. Ferric iron concentration between[3]:

- 0.1–0.2 g of Fe/m^3 is considered harmful to drip irrigation system,
- 0.2–1.5 g Fe/m^3 are having moderate, and
- 1.5 g Fe/m^3 are severely hazardous to drip emitters.

Normally, irrigation water containing 0.5 g iron is not suitable for drip irrigation system unless the drip lines are chemically treated. Regular chlorination can control the iron formation for pH <6.5.[18] Long-term use of high-level iron containing water is not suitable for drip irrigation. It is controlled by using sand filters to remove the oxidized iron precipitates. Injection of chlorine gas into the online system can also keep irrigation water free from iron precipitates with suitable filters.

12.7.2   WATER CONTAINING CA, MG

Water contains high levels of Ca, Mg, HCO_3, which leads to emitter clogging. Ca, Mg ions reacting with P fertilizers cause clogging in alkaline water leading to the formation of $CaCO_3$ precipitates that are accumulated in drip lines resulting in clogging. Scaling problems are commonly in water with pH >7.5 and bicarbonate content >5 mmol/L. Ca, Mg deposits increase with increasing pH of fertilizer solution.

12.7.3   INTERACTION AMONG CA, MG, AND FE IN FERTIGATION

Calcium concentration with low pH of fertigation water results in the precipitation of Ca–P and in some cases, iron also precipitates in the presence of P at pH>4 for iron and >5.5 for calcium. Therefore, P should not be used in drip irrigation if iron is present in irrigation water. In subsurface drip irrigation, irrigation clogging is more severe, which cannot be seen as in the surface drip irrigation. Phosphorus fertilizers are corrosive when they react with metals of tanks forming iron phosphate.[14,15] This can be cleaned by rinsing the fertigation system with HNO_3 to dissolve the precipitates present in the laterals and for adequate flow of irrigation water without any interruption. Because of such clogging problems, fertigation of P should be carefully monitored to avoid blockage of emitters and filters. The use of polyphosphate fertilizers causes gel formation with Ca, Mg thus blocking the emitters. To prevent gel formation, ammonium polyphosphate is recommended.

12.8   SUMMARY

This chapter focuses on application of soluble fertilizers through drip irrigation and performance of horticultural crops. By adopting this technique, the yield has improved by 20–30%. The fertilizer use efficiencies and water use efficiencies have been enhanced due to fertigation.

KEYWORDS

- **backflow**
- **bicarbonate**
- **crop root zone**
- **fertigation design**
- **laterals**

REFERENCES

1. Clark, G. A.; Smajstrla, A. G.; Haman, D. Z.; Zazueta. *Injection of Chemicals into Nitrogen System. Rates, Volumes and Injection Periods.* Fla. Coop. Ext. Bull 250; University of Florida, Gainesville, FL, 1990; p 12.

2. Drip Irrigation – per Drop More Crop. Unpublished Report NCPAH/2016/3; Indian Institute of Technology, Kharagpur, 2016; p 3.

3. Drip Fertigation in Horticultural Crops; 2016; www.crida.in/GSK.pdf (accessed Oct 15, 2019).

4. Enciso, J.; Porter, D. *Basics of Micro Irrigation.* Bulletin B-6160; Texas Cooperative Extension, Texas A&M University, Weslaco, TX, 2005; p 12.

5. Ford, H. W. *Iron Ochre and Related Sludge Deposits in Subsurface Drain Lines.* Circular 671; Florida Cooperative Extension Service, I. F. A. S, University of Florida, Gainesville, FL, 1982; p 8.

6. Haynes, R. J. Principles of Fertilizer Use for Trickle Irrigated Crops. *Lincoln—Canterbury (New Zealand): Ministry of Agriculture and Fisheries, Agricultural Research Division* **1989,** *6,* 235–255.

7. Hebbar, S. S.; Ramchandrappa, B. K.; Nanjappa, H. V.; Prabakar, M. Studies on NPK Drip Fertigation in field grown tomato (*Lycopersicon esculentum* Mill.). *Eur. J. Agro.* **2004,** *21,* 117–127.

8. Howell, T. A.; Hiler, E. A. 1974. Trickle Irrigation Lateral Design. *Trans. ASAE* **1974,** *17,* 902–908.

9. https://en.engormix.com/MA-agriculture/news/fertigation-fertilizer-sourcest3141/p0.htm (accessed Oct 13, 2019).

10. Kafkafi, U.; Tarchitzky. J. *Fertigation a Tool for Efficient Fertilizer and Water Management*; International Potash Institute (IPI), Paris, France, 2011; p 218.

11. Lawrence, J. S. *Fertigation*; fruitsandnuts.ucdavis.edu/files/73697.pdf (accessed Oct 13, 2019).

12. Locasico, S. J.; Smajstrla, A. G. Drip Irrigated Tomato as Affected by Water Quantity and n and k Application Timing. *Proc. Fla State Hortic. Soc.* **1989,** *102,* 307–309.

13. Magsen, H. In *Potassium Chloride in Fertigation*, Unpublished Paper Presented at 7th International Conference on Water and Irrigation; Tel Aviv, Israel; 13–16 May, 2002; p 8.

14. Malchi, I. Iron in Irrigation Water. *Hassadeh* (The Hebrew University of Jerusalem) **1986,** *66,* 1–16.

15. Malchi, I. *Personal and Internal Information*; Netafim Irrigation Inc., Fresno, CA, 1986.

16. Marschner, H. *Mineral Nutrition of Higher Plants*, 2nd ed., Vol. 78; Academic Press, London, 2005; pp 527–528.

17. Michael, A. M. *Irrigation Theory and Practice*, 2nd ed.; Vikas Publishing House, Noida, UP, India, 2010; p 772.

18. Nakayama, F. S.; Bucks, D. A. *Trickle Irrigation for Crop Production*; Elsevier Science Publishers B. V., New York, 1986; p 383.

19. Papadopoulos, I.; Eliades, G. A. Fertigation System for Experimental Purposes. *Plant Soil* **1987,** *1987,* 141–143.

20. Phene, C. J.; Beale, O. W. High Frequency Irrigation for Water Management in Humid Regions. *Am. J. Soil Sci.* **1976,** *40* (3), 430–436.

21. www.jains.com (accessed Oct 11, 2019).

22. www.ncpahindia.com/articles/article17.pdf (accessed Oct 11, 2019).

CHAPTER 13

HYDRAULIC PERFORMANCE OF DRIP FERTIGATION EQUIPMENT

E. K. KURIEN, NADIYA NESTHAD, ANU VARGHESE, and
E. K. MATHEW

ABSTRACT

Fertigation technology has become an integral part of drip irrigation system, where fertilizers are applied to crops along with irrigation. Venturi injector, dosmatic fertigation unit, and fertilizer tanks are used to incorporate fertilizers to the drip irrigation, and these were tested to study their hydraulic performance in this chapter. The suction rates and motive flow rates varied directly with the pressure drop in the system. Venturi injector had a superior suction rate compared to the dosmatic fertigation unit. Venturi injectors performed satisfactorily at system discharge rate >14.6 L/min. Dosmatic fertigation units were effective for discharge rate >1.1 L/min.

13.1 INTRODUCTION

The micro irrigation method delivers water directly near the root zone. The system delivers a constant rate of discharge that does not change significantly in the field. Micro irrigation has been adopted due to its various advantages, such as water saving, reduction in labor cost, better crop yield, high irrigation efficiency, and benefit–cost ratio. Judicious application of fertilizers and plant nutrients can enhance the system efficiency and ultimately the crop yield.

Fertigation is the method by which fertilizers are added to the irrigation water and applied to the crops through drip or similar micro irrigation system. Fertigation is gaining acceptance among the farmers due to its various advantages. The success of fertigation technology in terms of improved production depends on how efficiently crops take up the nutrients. Fertigation provides nitrogen, phosphorous, potassium, and other essential nutrients directly into the soil near the active root zone. This helps us to minimize the loss of nutrients and helps in increasing productivity and quality of produce. In order to reduce the water loss and weed infestation, crops are raised under fertigation system with the application of suitable mulching. Specialized equipment or applicators are needed for introducing fertilizers into the irrigation system. The applicators are to be incorporated into the system taking into account its hydraulic performance.

This chapter describes various drip fertigation equipment and their hydraulic performances.

13.2   DEVELOPMENTS IN DRIP IRRIGATION AND FERTIGATION TECHNOLOGY

13.2.1   DRIP IRRIGATION

Micro irrigation is the application of the water precisely on or below the soil surface by using surface or subsurface drip irrigation or bubbler or micro-sprinkler irrigation. Water can be applied either continuously or discontinuously near the plant so that the entire root system gets enough water. Water application is affected through drippers or emitters. In certain cases, water is applied in a spray pattern using micro-sprinklers or micro-jets.[5,8,9]

Drip irrigation is the most widely accepted micro irrigation method and has been practiced by farmers in an efficient manner in many countries. Over the past 20 years in India, area under drip irrigation has increased at least six folds to 10.3 million ha. Drip irrigation was commercialized in the 1950s by Israel engineers, and it reduces the volume of water applied to field up to 70%, and the crop yield is increased by 20–90%. Farmers have adopted this method mainly for fruits, vegetables, and high-value crops that can provide good economic returns on the investment.

In drip irrigation, water is pumped from the source, is filtered and passed through a pipe network under pressure, and delivered to the plants near the base of plants at slow rates.[10] The emitters attached to the lateral pipes allow irrigation water to drip out near the plant base. The rate of application of water must ensure the water need of the crop. The emitters or drippers should be of good quality, and their performance determines the quality of irrigation and success of the micro irrigation system. Automation technology has been widely introduced into the drip irrigation system to achieve optimal use of water resources. Automation helps in the precise application of irrigation water.[5] Automation is achieved by using suitable sensors and control systems.

Solar-powered wireless acquisition stations are employed for data acquisition and control of valves in an automated micro irrigation system.[3,6,7]

13.2.2 FERTIGATION

The fertigation system permits direct application of various fertilizer formulations to the active root zone. The major advantages of fertigation with drip irrigation are saving of water and labor, uniform distribution, and less damage to crop and soil, and thus resulting in higher yield. This method helps in the application of water-soluble fertilizers and other nutrients precisely to the soil at proper time in desired concentration.[4] Irrigation systems must be designed correctly to operate efficiently and to carry and deliver the fertilizers and nutrients to the soil. The fertilizers added to the system must be completely water soluble without leaving any residues. There should be a constant rate and pressure from main line for delivering nutrient solution.[8]

Fertigation is widely practiced in greenhouses and in protected cultivation. Several factors (such as plant species, humidity, and water availability within the greenhouse) affect the absorption and utilization of nutrients. Therefore, care and proper management of the media and appropriate fertigation program is essential for getting sustained productivity of crops in greenhouse cultivation. Excessive or imbalanced application of nutrients will result in improper plant growth.[7,9] Fertigation system is becoming more popular because of its advantages like higher fertilizer use efficiency, increased availability of nutrients to the plant. Regular supply of crop nutrients as and when required can be assured in a fertigation system. The 20–40% of fertilizer can be saved along with labor and energy savings. The fertigation equipment can be used for applying chemicals other than fertilizers for specific purposes.[2,8] It was reported that fertigation system is effective for commercial cultivation of horticultural and high-value crops.[1,6] The yield and quality of the produce were better on the adoption of this technology. Fertigation helps in sustaining soil health for better productivity. Fertigation systems can be automated and incorporated in automated greenhouse cultivation.

13.3 FERTIGATION EQUIPMENT

Water-soluble fertilizers are injected into the drip irrigation system using suitable fertilizer injectors, such as venturi devices/injectors, fertilizer bypass tank, and dosmatic injectors or fertigation pumps.

13.3.1 VENTURI INJECTOR

Venturi injector (Fig. 13.1) is popular and ideal for small micro irrigation systems because of their simplicity and low cost, and because they do not require additional external power source. Venturi injection rate of this device is low and is dependent on the system pressure and discharge. This will tend to limit its use on large systems, where the fertigation requirement is necessary. Small centrifugal pumps incorporated in the system can be used to overcome the drawback of small-area coverage.

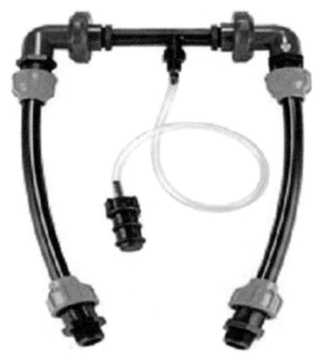

FIGURE 13.1   Venturi injector.

13.3.2 FERTILIZER TANK

Fertilizer tank (Fig. 13.2) operates on the principle of differential pressure created across the supply line with the help of a throttling value. This is a simple mild steel tank and is effective as a reliable injection device. It is used, where the higher concentration of the chemical or fertilizer does not harm the crop. The inlet and outlet of the fertilizer tank are connected to the main line at points, where a pressure difference exists. A pressure-reducing valve is also incorporated for the adequate operation. The chemical fertilizer dissolved in the water in the tank starts flowing through the pipeline, and the water through the pipeline replaces slowly the dissolved chemicals. Thus, the concentration of the applied chemical decreases continuously, and the chemical is gradually diluted until it has all been discharged fully into the irrigation system.

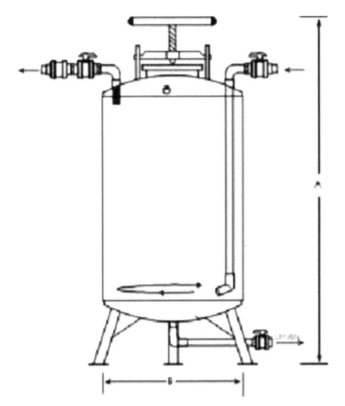

FIGURE 13.2 Fertilizer tank.

Fertilizer tanks can provide adequate service when their advantages and limitations are understood. They are frequently the best means of applying chemicals or fertilizers when no additional power source is available. The chief disadvantage is that the chemical concentration decreases with time. On small irrigation systems, the decreasing concentration may be unimportant and hence the use of fertigation tank for injection is ideal and simpler compared to the use of injection pump.

13.3.3   DOSMATIC INJECTOR OR FERTIGATION PUMP

These injectors or pumps are of piston or diaphragm type (Fig. 13.3) and are quite popular. They provide better precision than other methods of fertigation. However, they are costly and require frequent and high maintenance than the other fertigation methods. The hydraulic pressure of flowing water in the irrigation system makes the pump to operate. With a good injection pump, injection can be regulated precisely, and chemicals can be injected at constant concentration until the required amount has been applied.

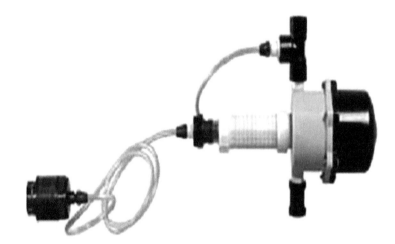

FIGURE 13.3   Dosmatic fertigation unit.

13.3.4   PERFORMANCE EVALUATION OF DIFFERENT FERTIGATION EQUIPMENT

13.3.4.1   VARIATION OF SUCTION RATE WITH PRESSURE DIFFERENCE

Three different types of fertigation equipment were tested for their performance and suitability for field application. An increase in suction rate was observed in the fertigation equipment with increased pressure differences of 0.1, 0.2, and 0.3 kg/cm^2. In the case of dosmatic fertigation unit, suction rate was 0.046 L/min at a pressure difference of 0.2 kg/cm^2. A value of suction rate of 0.103 L/min was observed in the case of venturi injector. At 0.8 kg/cm^2 of the pressure difference, corresponding suction rates for dosmatic fertigation unit and venturi injector were 0.163 and 0.23 L/min. The suction rate of venturi injector was higher than the dosmatic fertigation unit. The variation of suction rates with pressure difference for venturi injector and dosmatic fertigation unit is shown in Figure 13.4.

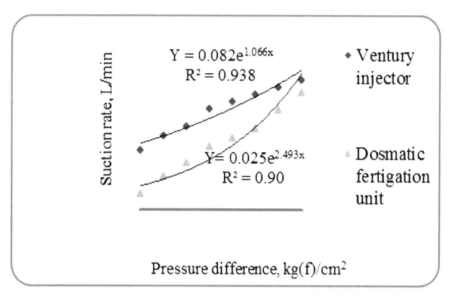

FIGURE 13.4 Variation of suction rate with pressure difference for venturi injector and dosmatic fertigation unit.

13.3.4.2 VARIATION OF FLOW RATES WITH PRESSURE DIFFERENCE

The variation in the flow rates is due to the change in the operating pressure, that is, the observed pressure difference. At low pressure, the flow through the bypass (i.e., the line connected with the venturi) was minimum. The increase in motive flow rate for venturi injector was 46% that was higher than that for dosmatic fertigation unit (45%) and fertilizer tank (45%) for the pressure difference between 0.1 and 0.8 kg/cm^2.

Table 13.1 shows the variation of motive flow rates with pressure difference for different fertigation equipment. For the minimum pressure difference of 0.1 kg/cm^2, the motive flow rates for venturi injector, dosmatic fertigation, and fertilizer tank was 14.6, 1.1, and 6.6 L/min, respectively. Therefore, venturi injector can be used only if the discharge rate is >14.6 L/min. Dosmatic fertigation unit and fertilizer tank can be used if the discharge rates are above 1.1 and 6.6 L/min.

TABLE 13.1   Variation in Flow Rates.

Pressure difference	Flow rates		
	Venturi	**Dosmatic**	**Fertilizer tank**
kg/cm^2		L/min	
0.1	14.60	1.10	6.60
0.2	17.60	1.32	7.92
0.3	19.40	1.45	8.74
0.4	20.90	1.56	9.40
0.5	23.76	1.78	10.60
0.6	23.50	1.92	11.55
0.7	26.54	1.99	11.90
0.8	27.13	2.03	12.29

13.4   HYDRAULIC PERFORMANCE OF THE DRIP IRRIGATION SYSTEM

13.4.1   VARIATION IN MANUFACTURING COEFFICIENT OF EMITTER WITH OPERATING PRESSURE

As the operating pressure is increased, the manufacturing coefficient of emitter is also increased. Variation in manufacturing coefficient of emitter with operating pressure is shown in Figure 13.5. For an operating pressure of 0.7 kg/cm^2, the variation in manufacturing coefficient of emitter was 17.8%. For an operating pressure of 0.5 kg/cm^2, the Cv (coefficient of variation) was 10.1%, which is acceptable showing good performance. This is in agreement with the findings of other authors, who observed that the drippers had the Cv value less than 5% indicating the good performance.[3,11]

13.4.2   VARIATION IN EMITTER FLOW

The emitter flow variation in dosmatic fertigation unit was decreased from 25 to 10% at various operating pressures. The decrease in emitter flow variation was 60% at operating pressures from 0.1 to 0.5 kg/cm². Variation of emitter flow with operating pressure is shown in Figure 13.6.

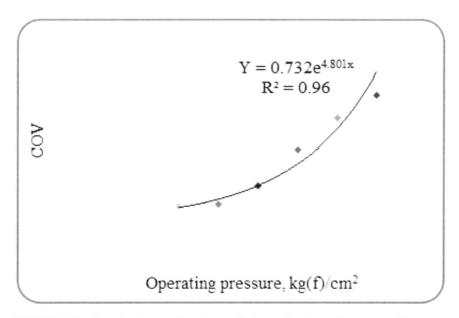

FIGURE 13.5   Variation in manufacturing coefficients of emitter with pressure difference.

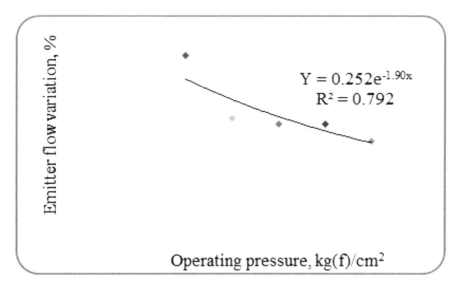

FIGURE 13.6   Variation in emitter flow with operating pressure.

13.4.3   EMISSION UNIFORMITY

Emission uniformity of the drip irrigation system was studied for different lateral lines of the system. For the first lateral line, the uniformity coefficient was 98% at an operating pressure of 1.2 kg/cm², while the same was 94% at an operating pressure of 1.2 kg/cm², for the second lateral line. This is in agreement with results by other researchers,[1,3] who reported that the inline drip irrigation system recorded >93% uniformity of water distribution. Variation in emission uniformity with operating pressure is shown in Figure 13.7. At low operating pressure, the discharge from the emitters was reduced. Similar observations were reported by Shindhe.[11]

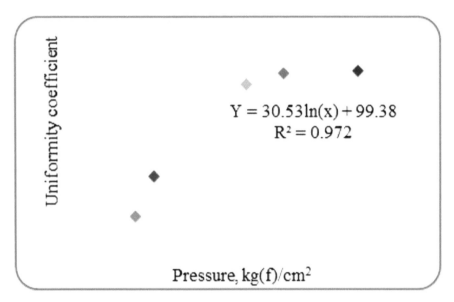

$$Y = 30.53\ln(x) + 99.38$$
$$R^2 = 0.972$$

FIGURE 13.7   Variation in uniformity coefficients.

13.5   SUMMARY

The venturi injector, dosmatic fertigation unit, and fertilizer tank were evaluated. The suction rates and motive flow rates varied directly with the pressure drop in these fertigation equipment. The manufacturing coefficient of variation of emitter was increased, and emitter flow variation was decreased with increase in operating pressure. The uniformity coefficient of the drip fertigation system was 98% for an operating pressure of 1.2 kg/cm². The farmers have a tendency to opt for venturi injectors in small farms.

KEYWORDS

- **dosmatic injector**
- **drip irrigation**
- **emitter flow variation**
- **fertigation**
- **fertilizer tank**

REFERENCES

1. Brain, J. Distribution Pattern of Micro Irrigation Spinner and Spray Emitters. *Appl. Eng. Agric.* **1989,** *5* (1), 50–56.
2. Khan, M. M.; Shivashankar, K.; Krishna Manohar, R. Fertigation in Horticultural Crops. ICAR Summer Short Course on Advances in Micro Irrigation and Fertigation; Univ. Agric. Sci., Dharward. In *Micro Irrigation*; Singh, H. P., Kaushish, S. P., Ashwani, K., Murthy, T. S., Samuel, J. C., Eds.; Central Board of Irrigation and Power, Government of India: New Delhi, 2001; p 79.
3. Kishor, L.; Reddy, K. S.; Sahu, R. K.; Bhanbarkar, D. M. *Studies on Hydraulic Characteristics of Drippers*; Paper Presentation at ISAE Convention; ISAE: New Delhi, 2005; p 42.
4. Kumar, A. Fertigation through Drip Irrigation. In ICAR Summer Short Course on Advances in Micro Irrigation and Fertigation; Univ. Agric. Sci., Dharward. In *Micro Irrigation*; Singh, H. P., Kaushish, S. P., Ashwani, K., Murthy, T. S., Samuel, J. C., Eds.; Central Board of Irrigation and Power, Government of India: New Delhi, 2001; pp 349–356.
5. Mahrin, D.; Semih, O. Optimizing the Inline Emitters for Higher Efficiency Silicon Solar Cells. *Sci. Res. Essays* **2011,** *6* (7), 1573–1582.
6. Manickasundaram, P. Principles and Practices of Fertigation. In *Short Course on Farming for the Future: Ecological and Economic Issues and Strategies: Micro Irrigation*; Kandasamy, O. S., Velayudham, K., Ramasamy, S., Muthukrishnan, P., Dcasenapathy, P., Velayuthan, A., Eds.; Central Board of Irrigation and Power, Government of India: New Delhi, 2005; pp 257–262.
7. Mortvedt, J. J. Mineral Nutrition of Greenhouse Crops. Dahlia Gardening International Symposium of Fertigation and the Environment Held at Institute of Technology. In *Micro Irrigation*; Singh, H. P., Kaushish, S. P., Ashwani, K., Murthy, T. S., Samuel, J. C., Eds.; Central Board of Irrigation and Power, Government of India: New Delhi, 1997; pp 110–118.
8. Nache-Gowda, V. Nutrition and Irrigation Management in Green House. National Training Program in Green House Technology, Held at UAS, Banglore, During 18–23 September. In *Micro Irrigation*; Singh, H. P., Kaushish, S. P., Ashwani K., Murthy, T. S., Samuel, J. C., Eds.; Central Board of Irrigation and Power, Government of India: New Delhi, 1996; pp 54–62.

9. Nakayama, F. S.; Bucks, D. A. (Eds.) *Trickle Irrigation for Crop Production: Design, Operation and Management*; Elsevier Publishers: Amsterdam, 2012; p 382.
10. Schwankl, L. J.; Hanson, L. R. Surface Drip Irrigation. In *Developments in Agricultural Engineering*; Lamm, F. R., Francis, S. N., Eds., 2007; pp 431–472.
11. Shindhe, P. P. In *Increasing Sugarcane Productivity Using Drip and Rain Gun Sprinkler Irrigation*, Proceedings International Conference on Plasticulture and Precision Farming, New Delhi; Choudhary, M. L., Chandra, P., Misra, P. Eds., 17–21 November, 2001; pp 201–206.

PART IV
Enhancement of Irrigation Efficiency

CHAPTER 14

MICRO IRRIGATION DEVELOPMENTS IN INDIA: TECHNO-ECONOMIC CHALLENGES

MANOJ P. SAMUEL and A. SURESH

ABSTRACT

Bringing more areas under micro irrigation (MI) is essential in India considering the issues with growing water scarcity, climate change, groundwater depletion with increasing demand for higher production of food, fodder, and fuel. Through MI, slow and steady application of water is possible directly to the root zone of plants. It helps in obtaining more crops per drop, reducing pressure on groundwater reserves and indirectly reduces the greenhouse gases emissions. MI system helps us to obtain high water use and application efficiency and can be integrated with fertigation and automation mechanisms that, in turn, will help in reducing the use of fertilizer and labor. Latest technologies like geographical information system, geographical position system, and hyper spectral imaging can be effectively used for precision agriculture practices employing MI systems. However, high initial cost and maintenance issues make farmers reluctant to adopt the technology on a large scale. The current focus of MI development strategy revolving around only on subsidy needs to be revisited so as to provide weightage to water conservation and recycling. It may be further linked to carbon credit concept and associated monetary benefits. In the context of climate change and water stress, MI programs are worthwhile adaptation and mitigation strategies.

14.1 INTRODUCTION

The giant leap in food production in India during the green revolution period is mainly attributed to the development of high yielding varieties and increased irrigated areas.[12] Micro irrigation (MI) has been adopted successfully in India especially for high-value crops. However, only less than 15% of the potential area has been brought under MI.[14]

MI has been successfully implemented in many places as water saving irrigation technology, and the advantage of the possibility of "more crop from drop" has drawn the interests of policy-makers to this novel system.[4] The necessity for MI is pronounced in the wake of extensive groundwater extraction for irrigation, climate change–induced rainfall variability necessitating enhancement of water use efficiency. In future, judicious use of available water resources would emerge as the critical factor for the growth of an agricultural sector, including that of high-value horticulture crops, livestock, and fisheries sectors depend on the water availability. Therefore, future strategy for agricultural growth needs to concentrate on water budgeting, thus considering demand from all sectors.

It is critical for the ambitious crop diversification scientist in agricultural sector. MI helps us to realize higher value of output per unit amount of input water through efficient usage. In this context, water mapping, water budgeting, and MI emerge critical factors in developing location-specific agricultural plans and investment decisions. The approach to MI, in this background, needs to acquire the status of a policy tool.

Government of India (GoI) and State Governments in India promote MI by pumping subsidy schemes and giving other financial incentives.[9,15] However, many times it is noticed that the subsidy schemes fail because of the lack of ownership feeling, poor maintenance and follow-up, and poor identification of beneficiaries. The technical and economic constraints attached with the MI system also have to be addressed. However, MI has to be promoted as an efficient and effective replacement of traditional irrigation system, considering the climate change and water scarcity issues.

The environmental issues (like increasing emissions and overexploitation of groundwater) may also be linked to irrigation strategies, and a paradigm shift toward MI would, in turn, reduce the intensity of the said issues. As MI systems require comparatively lower power, therefore, there is a possibility of integrating solar and other renewable energy power sources in the system so that it can function as a stand-alone.

This chapter discusses various issues related to MI, the bottlenecks in the adoption of this technology among the farmers, and the policy implications by GoI.

14.2   MICRO IRRIGATION

MI helps efficient use of water, realizing higher value of output per drop of water.[8,19] Slow and steady application of water directly to the root zone of plants is possible through drip irrigation system that consists of a network of economically designed plastic pipes and emitters. It has high efficiency and effectiveness of water application as shown in Table 14.1. The reduced loss of water during conveyance, coupled with high application efficiency and low evaporation losses, makes MI a highly efficient irrigation system.[6]

TABLE 14.1   Irrigation Efficiencies of Different Irrigation Systems.

Parameter	Irrigation efficiency (%)		
	Surface	**Sprinkler**	**Micro/Drip**
Application	40–70	60–80	90–95
Conveyance	40–50	90–100	90–100
Evaporation losses	30–40	30–40	20–25
Overall efficiency	30–35	50–70	80–90

The concept of precision farming is essentially related to the MI to suit the needs for precise application of agricultural inputs, including water. MI systems are suitable for arid regions, subhumid and humid zones, where water supplies are limited or water is expensive. It has a unique role in urban and *peri*-urban agriculture, where water supply is limited and the focus is on the cultivation of fruits and vegetables. In irrigated agriculture, MI is used extensively in orchards, gardens, furrow crops, mulched crops, greenhouses, and nurseries. In urban landscapes, MI is widely used with ornamental plantings, terrace, and protected farming.

The development of MI in the world dates back to experiments in Germany in 1860s, followed by studies in the United States in 1913, Germany in 1920, Israel in 1940, the United Kingdom in 1948, and Israel again in 1960 and later in 1969. As an important breakthrough, in 1920 perforated pipe drip irrigation was introduced in Germany. Israel started selling drip irrigation pipes on commercial basis on the global market in 1969.[10]

Major components of an MI system include emitters, micro-tubing, controller/timer, backflow preventer, valves, filter, pressure regulator, pipes, flush valve/cap, etc. The emitter, which is a metering device made from plastic, delivers a small but precise discharge. These emitters dissipate water pressure through the use of long paths, small orifices, or diaphragms. Specially made emitters for undulated areas (including pressure-compensating types) can discharge water at a constant rate over a wide range of pressures. There may be three types of emitters: drip, bubbler, and micro-sprinkler.

In the drip mode, water is applied as droplets or trickles, while in a bubbler water "bubbles out" from the emitter. However, the irrigation water is released as a fine spray or mist in the micro-sprinkler. The discharge capacity of the emitters differs, such as 2, 4, 8, 10, 16, and 32 lph. Some emitters are designed as online emitters to apply water to closely spaced crops in rows.

Even in years with above average rainfall, crop growth and yields in many parts of India can still be limited by inadequate water in the soil (soil water) at critical periods during the growing season. During summer months, the soil–water deficit is high and may be detrimental to the plants if the dry spell extends. Supplementing rainfall with irrigation during these periods not only can increase crop yield but also can improve the quality of fruits. This is especially important for high-value crops like flowers, vegetables, and fruits.

Agronomic crops can also benefit from irrigation, but potential returns may not be enough to justify the cost. The economics of irrigation also depend on market policies. Depending on the crop, adequate soil water is most critical during the stages of seed germination, seedling or transplanting, flowering, fruit-set, and fruit enlargement. Fluctuating water supply may lead to the development of water-related physiological disorders too.

The selection of an appropriate irrigation system depends on many factors, such as available capital, cropping system, crop rotations, equipment, labor, soil type, soil topography water quantity and quality, power requirements, and total area irrigated. Under Indian context, plantation crops like coconut, arecanut, citrus, vegetables, fruit crops (like bananas, papaya, guava, and grapes) respond well to drip irrigation. Online emitters are suited for closely spaced crops, whereas point source emitters are suited to wider spaced crops, such as fruit trees and vineyards. The major MI systems include drip, micro-sprinkler, mist, fogger, and bubbler.

14.2.1 DRIP IRRIGATION

In drip irrigation system, water is applied directly to the plant at right time, right location, and in the correct quantity. Generally, water is applied drop by drop directly to the root zone of the plant by means of emitters or drippers. Emitters are fitted to lateral pipe lines made up of preferably linear low-density polyethylene. For close growing crops (e.g., sugar cane), built-in emitters in the lateral line are used. The slow, steady, and regular application of controlled water gradually develops an onion-shaped wetting zone in the root zone of the plant and thus ensures efficient irrigation. The application efficiency of drip irrigation is as high as 90%.[19]

Drip irrigation system has many advantages, such as reductions in water use, labor savings, energy/fuel saving, reduced incidence of foliar diseases, ease of fertigation, and ability to irrigate even during other field operations, such as spraying and harvesting. Main disadvantage of drip irrigation is its high initial investment cost. Moreover, compared to other irrigation methods, drip system needs a filtration unit to prevent clogging of emitters.

14.2.2 BUBBLER IRRIGATION

Bubblers release water from the emitter in an umbrella pattern. The diaphragms and small orifices in the bubbler help it to act as a pressure-compensating emitter. Discharge capacity of a bubbler generally varies from 10 to 100 lph. Bubblers apply water in a uniform pattern, because these are less susceptible to clogging problems.

14.2.3 MICRO-SPRINKLER IRRIGATION

Micro-sprinklers release water in the form of fine spray, mist, or fog. They can spread water around the plant in a predetermined pattern. They are generally connected to the lateral by means of micro-tubes. A support stake is also used to keep the micro-sprinkler head erect and straight. Their discharge capacity varies from 10 to 150 lph and can cover larger areas.

14.3 ADVANCES IN MICRO IRRIGATION TECHNOLOGY

Irrigation technology is fast changing throughout the world. Progress in Israel is worth mentioning in the development of high-end technologies for water collection, conservation, precision application, and recycling in field and protected conditions. In the irrigation sector, sensor-aided controllers in automatic systems can regulate the application of water and nutrients. These are gaining popularity not only in climate-controlled poly-houses for export-oriented high-value crops, but also in the field crops. In view of enhanced water saving, the policy needs to target add-on features, including precision water management systems.

MI is an integral component of precision agriculture technology that has been slowly but steadily getting popular among high-value crops. Integration of geographical information system (GIS), geographical position system (GPS), and hyperspectral imaging with precise application of water and fertilizer through MI systems creates much high-value proposition.

MI is fast moving toward automation and precision farming, which help one to reduce the labor cost and use of resource inputs, such as water, nutrients, and fertilizers. Automation and precision farming are preemptive responses to a declining labor force and high wage rates. With automatic systems, sensors gather information on moisture levels of plants, soil, or atmosphere and switch on/off according to the input goals that are preset. Automated systems virtually eliminate the need to monitor plant nutrition levels and elevate the agronomic management regime in a different plane.

The advanced tools and systems, which can be integrated to an MI system, are indicated in Figure 14.1.

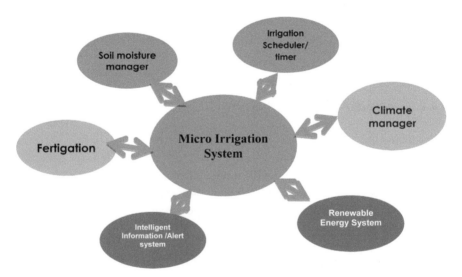

FIGURE 14.1　Integrated micro irrigation system.

14.3.1　FERTIGATION

Fertigation is the process of applying water-soluble fertilizers to the plants through MI system. This helps one to reduce the fertilizer use and ensure more efficient application. The fertilizers can be injected into the drip lines by means of a fertilizer tank or a venturi system. The venturi device attached to the main line helps to cause a vacuum, which, in turn, induces the suction of fertilizer solution into the line. Using the fertigation system, nutrients and water can be supplied to the crop with high precision, in both spatial and temporal manner, thereby allowing high fertilizer application efficiency.[8]

14.3.2 AUTOMATION

Automatic irrigation system allows efficient use of water resources. Automatic systems help in applying water precisely at right time and help one to reduce labor cost. In an automatic irrigation system, automatic valves (such as solenoid valves) are used, which are controlled by electrical signals. These valves can control the flow of water to different parts of the field by opening and closing due to an electrical signal from the controller. The system can be operated either using some predefined algorithms based on crop water requirements or by sensing the real-time soil moisture values. Apart from electric solenoid valves, one will need a time clock, an automatic controller unit, and/or preprogramed master control units.

14.4 DEVELOPMENT ISSUES IN MICRO IRRIGATION: INDIA

MI development has been supported with high subsidy, envisaged through different schemes of Central and State Governments in India. There was about 8.7 million ha of cultivated land under MI in 2017. However, this constitutes only 13% of the 70 million ha of potential area that can be brought under MI.[5] Andhra Pradesh, Rajasthan, Gujarat, Haryana, Madhya Pradesh, Maharashtra, Punjab, Tamil Nadu, Telangana, and Uttar Pradesh account for 80% of the area under MI (Table 14.2).

In the case of drip irrigation, Maharashtra, undivided Andhra Pradesh (Andhra Pradesh and Telangana), Karnataka, and Gujarat together constitute about 85%.

Rajasthan and Haryana are on the top of the list for sprinklers (Table 14.2). Some high-potential states (such as Madhya Pradesh, Punjab, and Haryana) fare low on the adoption of MI. However, these states fall in the realm of unsustainable extraction of irrigation water from the tube wells. It warrants having a deep pondering over the patchy spread of MI. In this context, Global Green Growth Institute (GGGI) indicated that the massive promotional activities of the manufacturing companies and easy physical availability of equipment have significantly influenced the adoption.[3] For example, in

Jalgaon district, MI has significant reach due to the presence of Jain Irrigation System (a major manufacturer of MI systems in India).

TABLE 14.2 Spread of Micro irrigation in India: 2016.

State	Area ('000 ha)	Share to all India (%)	Share to the state (%)	
			Drip	**Sprinkler**
Andhra Pradesh	1323.21	15.34	71.99	28.01
Bihar	107.92	1.25	9.04	90.96
Chhattisgarh	271.15	3.14	6.67	93.33
Gujarat	1068.81	12.39	50.01	49.99
Haryana	576.83	6.69	4.27	95.73
Himachal Pradesh	7.82	0.09	54.77	45.23
Jharkhand	20.75	0.24	52.20	47.80
Karnataka	953.35	11.05	51.08	48.92
Kerala	30.32	0.35	75.46	24.54
Madhya Pradesh	430.66	4.99	52.12	47.88
Maharashtra	1309.67	15.18	70.59	29.41
Odisha	104.84	1.22	18.49	81.51
Punjab	47.09	0.55	73.65	26.35
Rajasthan	1752.67	20.32	11.62	88.38
Sikkim	9.09	0.11	66.52	33.48
Tamil Nadu	363.36	4.21	90.38	9.62
Telangana	94.97	1.10	79.82	20.18
Uttar Pradesh	42.66	0.49	39.40	60.60
Uttarakhand	1.01	0.01	68.77	31.23
West Bengal	51.18	0.59	1.18	98.82
All India: Total	8626.78	100.00	45.44	54.56

Source: Adapted from Ref. [5].

Based on net-cropped area, the average penetration level of MI is only 5.5%, which is much lower compared to other countries. Only few states in India have the penetration level higher than the national average in India. The major states with high penetration level are Haryana (16.3%), Andhra Pradesh (10.4%), Rajasthan (9.3%), Karnataka (8.5%), Gujarat (8.1%), Maharashtra (7.3%), and Tamil Nadu (6.4%). According to the report by Irrigation Association of India][1,7], a steady growth rate of 9.6% has been observed in MI during 2005–2015, and recently sprinkler irrigation has also shown the highest adoption rate compared to drip irrigation.

14.5   NEW APPROACHES IN MICRO IRRIGATION

14.5.1   MICRO IRRIGATION IN CANAL COMMANDS

MI has largely been developed with minor irrigation systems. In view of water scarcity and better utilization of available water, conjunctive use of surface and ground water is being promoted. It has resulted in high water productivity and water saving. However, conjunctive use of water was not popular in canal commands. The recent experience suggests the possibility of integrating MI in canal commands (e.g., in Rajasthan, Karnataka, Gujarat, Odisha, and Tamil Nadu).

Generally, flood irrigation using gravity flow is being followed in canal commands. The release of water from canal commands suffers due to the lack of control by farmers, untimely operations, and inappropriate release of quantity of water, etc. Many of these constraints can be overcome by collecting water in certain storage structures, which can be used at an appropriate manner later. This collected water can be released through MI systems. This practice is already in vogue in some states, such as Gujarat and Rajasthan. This technology can be replicated in other locations in India, but it may require redesigning the existing irrigation systems.

Another strategy is to redesign the cropping pattern. Cultivation of high water-consuming crops in canal commands needs to be reconsidered, without which the inequality in water access would remain a serious issue. The farmers need to be incentivized to conserve water at the basin level, which can be later used for crop cultivation through MI. The water conservation can be achieved through the adoption of MI along with gradual withdrawal from flood irrigation methods. Water budgeting at basin level is the first step toward this. This along with land use and crop planning can effectively alter water demands and can lead to achieve the target of more crop per drop.

14.5.2   LEVERAGE FOR RAINWATER HARVESTING AND WATER RECYCLING

The MI implementation in India has remained a stand-alone target without considering efforts to improve the water supply. Recharging the groundwater is of critical importance to achieve long-term targets to improve water productivity and intensive agriculture. Further, water recharge has significant role in ensuring water availability in peninsular rivers. This can be achieved by using engineering and associated source protection practices involving communities/stakeholders.

The diversion of surface water storage for drinking water purpose is a major reason for the decline in the water availability for agricultural purpose. Rainwater harvesting is an alternative for conventional water supply schemes and it can be popularized.[17] Along with efforts to harvest rainwater, the sewage water can be treated and recycled for agricultural use through MI. However, in India, only about 20–30% of about 40–60 billion liters of sewage are treated.

14.5.3   MICRO IRRIGATION DEVELOPMENT AND POLICY

The GoI has accorded a priority for MI technology. In 2006, the Centrally Sponsored Scheme for MI was initiated to increase the acreage under MI in India. It was enhanced into a mission mode in 2010 and rechristened as National Mission on MI (NMMI). In the wake of India's action plan in tune with climate challenges, National Mission on Sustainable Agriculture was initiated, and NMMI was brought under its ambit. The major component of the program is to promote MI by providing capital subsidy.

In 2015, *Prime Minister's Krishi Sinchayee Yojana* was initiated, with focus on the entire supply chain. Other programs on MI were merged into this scheme. The funding allocation under these schemes was enhanced over years. The "more crop per drop" is a major component of this scheme focusing on MI development. The GoI continued support to MI projects through budgetary allocations that was also reflected in the 2018 budget.

One of the deficiencies of MI policy is that it has been promoted as a stand-alone scheme, with minimal linkages with other dimensions of water management. Therefore, the MI schemes need to be linked with efforts in water conserving, recycling and reuse, and recharging the groundwater source.

14.6   CHALLENGES AND WAY AHEAD

According to a recent report by the Irrigation Association of India,[1] the major challenges faced by the MI sector in India are

- misplaced priority on MI;
- lack of departmental support and IT-backed operations;
- delay in implementation, release of guidelines/government orders, uncertainty, and sporadic changes in scheme guidelines;

- inefficient subsidy disbursement process; and
- few easy financing mechanisms for farmers.

14.6.1 TECHNICAL AND ECONOMIC CONSTRAINTS

The major technical constraints in MI are the damages occurring to pipes and distributaries and clogging of emitters; proper maintenance of the drip system, including fitters; and service for the equipment. However, it has been observed that the equipment service systems are inadequate, and farmers require technical training at least in the initial phases. Availability of quality maintenance services is significant in attracting farmers for adoption of MI. Establishing custom hiring centers in this direction can address this problem. Also, this would provide employment opportunities.[6]

Another important constraint is the high capital investment for installing MI system. Studies indicate to the existence of high demand for low-cost MI systems.[3] However, the government schemes warrant purchase of systems with certain quality levels that are generally priced higher. GGGI mentions existence of high demand for low-cost MI systems in many parts of India. This gives rise to the concept of affordable drip irrigation.

According to Indian Council of Agricultural Research, although MI is technically efficient, yet it is expensive in terms of cost per ha and it further depends on the size of farm and type of crop grown. The indicative cost of drip irrigation for closed-spaced crops varies from $316 to $6927/ha for crops with varied spacing. Therefore, to support the farmer on the huge capital costs, GoI has introduced many financial assistance schemes and subsidies to the extent of 40–90% of the cost of MI system.[7]

14.6.2 CLIMATIC AND ENVIRONMENTAL ISSUES

Climate change would impact the availability of surface and groundwater resources. Currently, the extraction of water is by using diesel or electricity, thereby emitting huge quantities of greenhouse gases (GHGs). High demand for water from irrigation and domestic sector would increase the demand for energy, which, in turn, would result in further augmentation of GHG emissions.

Groundwater pumping in India accounts for a significant proportion of carbon emissions. It is estimated that for every 1% increase in ground-water irrigation, GHG emission tends to increase by 2.2%. Furthermore, it is estimated that for every 1% improvement in irrigation efficiency, GHG emissions are reduced by 2.1%.[18] Agriculture contributes to about 20% of India's net GHG emissions annually.[16] Improving water use efficiency and moving toward renewable energy sources will reduce GHGs and contribute to climate change adaptation.

Therefore, the future policies on agricultural growth must embrace a sustainable and green economy model, with a reduction in energy, carbon, and water footprints. It is necessary for meeting India's commitments for limiting the GHG emissions. Under a business as usual scenario, the twin objectives of meeting irrigation water demand with reduced carbon footprints cannot be materialized. Hence, the water and energy savings will have to come from better demand management, and an improvement in water and energy efficiency is a must.

14.6.3 POLICY IMPLICATIONS

Government policies influence energy consumption and energy efficiency in irrigation systems. Tube well owners with subsidized electricity operate electric pumps for longer duration than warranted.[11] Highly subsidized power has reduced marginal-cost of water extraction, thereby disincentivizing farmers from adopting water conservation measures. Further, the water is applied to the field through flood irrigation with efficiency <40%.

Competition for limited water from aquifers results in competitive deepening of borewells, and consequent overextraction of groundwater, which has several social, economic, and ecological repercussions,[2,13] which include drying of aquifers and use of pump-sets with high HP leading to higher energy use and emissions. Bringing more area under MI network and gradual conversion of flood-irrigated lands to micro-irrigated ones would be helpful in reducing the pressure on groundwater reserves. It would also help in reducing the energy costs and carbon footprints. Policy framework may be readjusted by providing subsidized power only to the farmers ensuring high water application and water use efficiency.

Water budgeting may be given priority at basin level as it is a highly localized agricultural unit, taking into consideration the demand and supply position of water. MI can be an integral part of the crop plans and water budgets. Conjunctive use of water in canal commands needs to take into consideration the possibilities of MI as well.

The geographical locations can be prioritized for popularization of MI taking into consideration the regional crop planning and water budget. One important criterion may be the current stage of groundwater development, which could effectively function as an indicator of sustainability. The regions, which are classified "overextracted" and "critical," can be accorded top priority.

The high initial cost hinders wide adoption of MI. Considering this, subsidy scheme can be reframed in such a way to enhance the completion among traders and manufactures so that farmers could afford the system. The MI technology has to be integrated with latest technological developments in the field of precision agriculture, GIS, GPS, multispectral imaging, etc. Better options should be made available to the framers in maintaining the MI system in a hassle free manner.

14.7   SUMMARY

The MI system works on the principle of "more crop per drop." It increases the water use efficiency and is the most effective technology for irrigation in water deficit areas, difficult terrains, and in controlled atmospheres. It is preferred over the traditional system in developed countries because of the scope of automating the entire system. Fertilizers and pesticides can effectively be supplied to the plants through MI. Only about 13% of potential areas could be brought under MI, warranting a relook at the MI policy. Solar-powered MI systems can further reduce the carbon emission and have the potential to save electricity. The current strategy of MI development focusing only on subsidy needs to be revisited with an aim to accord due weightage for water conservation efforts and water recycling. Over a period of time, the subsidy schemes can be reformatted into incentive-based schemes.

KEYWORDS

- **canal commands**
- **GHG emissions**
- **groundwater**
- **micro irrigation**
- **rainwater harvesting**
- **water use efficiency**

REFERENCES

1. Anonymous. Accelerating Growth of Indian Agriculture: Micro Irrigation an Efficient Solution. In *Strategy Paper: Future Prospects of Micro Irrigation in India*; Irrigation Association of India (IAI), Federation of Indian Chambers of Commerce & Industry (FICCI) & Grant Thornton India, 2016; p 10.
2. Chandrakanth, M. G.; Arun, V. Externalities in Groundwater Irrigation in Hard Rock Areas. *Indian J. Agric. Econ.* **1997,** *52,* 761–771.
3. Global Green Growth Institute (GGGI). *Implementation Roadmap for Karnataka Micro Irrigation Policy*; GGGI (Global Green Growth Institute): Seoul, Republic of Korea, 2015; p 98.
4. Government of India (GoI). *Budget Highlights (Key Features)*, 2017. http://indiabudget. nic.in/ (accessed Aug 31, 2019).
5. Government of India (GoI). *Agricultural Statistics at a Glance*, 2017; Directorate of Economics and Statistics, Ministry of Agriculture and Farmers' Welfare, Government of India: New Delhi, 2017; p 216.
6. Goyal, M. R. *Management of Drip/Trickle or Micro Irrigation*; CRC Press: Boca Raton, FL, 2012; p 412.
7. Harsha, J. *Micro Irrigation in India: An Assessment of Bottlenecks and Realities*; Global Water Forum, 2017. http://www.globalwaterforum.org/2017/06/13/Microirrigation-in-india-an-assessment-of-bottlenecks-and-realities/ (accessed Aug 31, 2109).
8. Incrocci, L.; Massa, D.; Pardossi, A. New Trends in the Fertigation Management of Irrigated Vegetable Crops. *Horticulturae* **2017,** *3,* 8. doi:10.3390/horticulturae/3020037.
9. Indian Council of Agricultural Research (ICAR). *Sustainable Farm Income from Integrated Farming in Arid Regions*; ICAR: New Delhi, India, 2016. http://www.icar. org.in/en/node/5718.
10. Indian National Commission on Irrigation and Drainage. *Drip Irrigation in India*; Indian National Committee on Irrigation and Drainage: New Delhi, 1994. http://cwc.gov.in/ INCSW/assets/pdf/Drip%20Irrigation%20in%20India.pdf (accessed Mar 31, 2019).
11. International Water Management Institute. *The Energy-Irrigation Nexus*; International Water Management Institute (IWMI): Sri Lanka, 2003; p 98.
12. James, K. *The Key to Understanding Global History*; Jenson Books Inc.: North Logan, UT, 1998; p 78.
13. Kumar, M. D.; Scott, C. A.; Singh, O. P. Can India Raise Agricultural Productivity While Reducing Groundwater and Energy Use? *Int. J. Water Res. Dev.* **2013,** *29,* 557–573.
14. Livemint. *Micro Irrigation Lags Far Behind Potential, Shows Study*, 2016. https:// www.livemint.com/Politics/BkgERIIfG77UzWZRC3HsAM/Microirrigation-lags-far-behind-potential-shows-study.html (accessed Mar 31, 2019).
15. Malik, R. P. S.; Giordano, M.; Rathore, M. S. The Negative Impact of Subsidies on the Adoption of Drip Irrigation in India: Evidence from Madhya Pradesh. *Int. J. Water Res. Dev.* **2016,** *2016,* 1–12.
16. Nelson, G. C. *Greenhouse Gas Mitigation- Issues for Indian Agriculture*; Discussion Paper 00900; International food Policy Research Institute: New Delhi, 2009; p 76.
17. Samuel, M. P.; Satapahy, K. K. *Curr. Sci.*, **2008,** *95* (9), 10–15.
18. Shah, T. Climate Change and Groundwater: India's Opportunities for Mitigation and Adaptation. In *Water Resources Policies in South Asia*; Prakash, A., Singh, S., Goodrich,

C. G.; Janakarajan, S., Eds.; Routledge, New Delhi, India, 2013; pp 213–243. https://hdl.handle.net/10568/37229.

19. Sivanappan, R. K. Technologies for Water Harvesting and Soil Moisture Conservation in Small Watersheds for Small-scale Irrigation. In *FAO Irrigation Technology Transfer in Support of Food Security (Water Reports–14)*, Proceedings of a Sub-Regional Workshop, Harare, Zimbabwe, April 14–17, 1997. http://www.fao.org/3/w7314e/w7314e0q.htm (accessed Nov 30, 2019).

CHAPTER 15

DRIP IRRIGATION SYSTEMS FOR ENHANCING INPUT USE EFFICIENCY

K. V. RAMANA RAO and SUCHI GANGWAR

ABSTRACT

The drip irrigation system has increased the crop yield compared with conventional irrigation. Irrigation operation uses significant portion of energy in agricultural production. The total inputs energy (MJ/ha) for conventional and drip irrigation systems was estimated as capsicum (7650 and 3051), tomato (6120 and 3243), potato (7140 and 4569), okra (8160 and 4406), cucumber (5100 and 4233), chili (5100 and 4539), pea (4080 and 3100), respectively. The total energy output (MJ/ha) for conventional and drip irrigation systems was estimated as capsicum (1,456,000 and 3,648,000), tomato (1,600,000 and 3,840,000), potato (1,488,000 and 2,316,000), okra (1,028,800 and 1,912,800), cucumber (2,968,000 and 3,576,000), chili (159,200 and 219,200), pea (4,696,000 and 6,192,000), respectively. The drip fertigation resulted in higher crop yields by maintaining the favorable soil environment and higher uptake of major nutrients in vegetable crops compared with granular form of fertilizers. Although the initial investment for drip irrigation system is high, the operational costs for labor use is minimal compared with conventional irrigation. The maximum cost savings (%) with drip irrigation compared with conventional irrigation were possible in capsicum (25), tomato (25), potato (19), okra (13.8), cucumber (15.7), chili (21.8), and pea (19.23).

15.1 INTRODUCTION

The annual horticultural production in India is about 149 million tons. Due to the short duration, rich in nutrition, ability in generating employment opportunities, Government of India is promoting vegetable cultivation under various schemes. With the diversified agroclimatic conditions coupled with distinct seasons, different types of vegetables can be grown in India. India contributes about 14.5% of total vegetable production and ranks second in the world. Conventionally vegetable growers adopt furrow irrigation practices for irrigating the crops, which has 25–40% of application efficiency.

Micro irrigation has emerged as an appropriate water application technique for horticultural and agricultural crops. Drip irrigation is receiving better acceptance and adoption, particularly in areas of water scarcity. Therefore, the efforts are necessary to produce more "crop per drop" of water. Drip irrigation provides effective way to apply water and nutrients to the plant, resulting not only in saving of water but also enhanced crop yields.[2,3,8,15] Water saving in drip irrigation is due to minimization of conveyance and application losses and water losses through ponding, evaporation, seepage, and deep percolation.

Since the water is applied in small quantities, it is possible to maintain constant soil moisture tension throughout the growing period. Water saving up to 50% is also possible with drip irrigation.[5,7,18,20] In drip irrigation system, only 15–60% of soil surface is wetted and the wetting pattern is unsaturated. Experimental investigations have resulted in positive response in majority of the vegetable crops by providing more frequent irrigation under drip system. By providing more frequent irrigation, the salts accumulation in the root zone can also be managed effectively.

In fertigation, fertilizers are applied along with flowing water through drip system resulting in higher fertilizer use efficiency (FUE), which in turn enhances crop yield. Frequent applications of soluble nutrients save labor and fertilizer, reduce soil compaction, and enhance crop productivity. The required quantities of fertilizers are directly placed in the plant root zone during critical periods using fertigation.[9,13,17]

The traditional practice of broadcasting with high doses of inorganic fertilizers not only causes economic loss but also pollutes soil and environment. By adopting drip fertigation, the fertilizer requirement can be reduced by 15–25% without compromising the crop yield.

Drip irrigation along with plastic mulching is effective in increasing the yield compared with no mulching under surface irrigation or drip irrigation.[14,16,21] Mulching is used to cover soil surface around the plants to create congenial condition for the plant growth. Polyethylene mulches are widely used in the cultivation of vegetables, due to their ease in transport, availability, and installation. The adoption of mulching has desirable effects, such as weed control, temperature moderation, salinity reduction. It also exerts decisive effects on earliness, yield, and quality of the crop. Apart from weed control, application of mulch films along with drip irrigation can enhance the nutrient use efficiency of crops.[1,4,10,12]

This chapter focuses on water use efficiency (WUE), FUE, input use efficiency, crop productivity, and yield of vegetable crops under drip irrigation compared with conventional irrigation method.

15.2 WATER USE EFFICIENCY

With drip irrigation system, about 50–70% of water can be saved compared with conventional irrigation systems. Drip irrigation applies water more frequently that minimizes run time of pump sets. Several research studies (Table 15.1) have reported that the drip irrigation system increased the crop yield (t/ha), such as capsicum (45.6), tomato (67.0), potato (28.9), okra (23.9), cucumber (44.7), chili (2.7), and pea (77.5) compared with conventional irrigation.

A close perusal of Table 15.1 indicates that about 10–60% increase in yield is possible with drip irrigation compared with conventional method of irrigation. Also compared with conventional, drip irrigation resulted in higher WUE of capsicum [60.80 kg/(ha mm)], tomato [80 kg/(ha mm)], potato [41.36 kg/(ha mm)], okra [29.89 kg/(ha mm)], cucumber [89.40 kg/(ha mm)], chili [5.48 kg/(ha mm)], and pea [193.50 kg/(ha mm)].

The published literature indicated that additional amount of water under the conventional irrigation technique is being used in capsicum (4508 m^3/ha), tomato (2820 m^3/ha), potato (2520 m^3/ha), okra (3680 m^3/ha), cucumber (850 m^3/ha), chili (550 m^3/ha), and pea (173 m^3/ha) compared with drip irrigation method.

15.3 ENERGY USE EFFICIENCY

Water-energy productivity is the amount of yield that is obtained from a unit of energy (kWh) consumed during the irrigation. Among all other agricultural field operations, water pumping consumes several times more energy.[11]

Energy requirements for agricultural production are increased when the water conveyance and water application are inefficient. Furthermore, achievable energy conversion efficiency for lifting water is about 20% in developed parts of the world compared with 12.5% elsewhere. Therefore, in order to have more sustainable energy utilization, water application

TABLE 15.1 Irrigation Parameters under Conventional and Drip Irrigation Methods for Different Vegetable Crops.

Crop	Irrigation practice	Crop yield	Crop water use	Quantity of water applied	The total energy input	The total energy output	Water energy productivity	Water saving over conventional	Ref.
		kg/ha	mm	m3/ha	MJ/ha	MJ/ha	kg/kWh	%	
Capsicum	Conventional	18,200	750	7500	7650	14,56,000	13.5	60.1	[14]
	Drip	45,600	750	2992	3051.84	36,48,000	84.6		
Tomato	Conventional	20,000	600	6000	6120	16,00,000	18.5	58.33	[13]
	Drip	48,000	600	3180	3243.6	38,40,000	83.85		
Potato	Conventional	18,600	700	7000	7140	14,88,000	14.8	36.00	[4]
	Drip	28,950	700	4480	4569.6	23,16,000	35.9		
Okra	Conventional	12,860	800	8000	8160	10,28,800	8.9	46.00	[8]
	Drip	23,910	800	4320	4406.4	19,12,800	30.7		
Cucumber	Conventional	37,100	500	5000	5100	29,68,000	41.2	17.00	[19]
	Drip	44,700	500	4150	4233	35,76,000	59.8		
Chili	Conventional	1990	500	5000	5100	1,59,200	2.21	27.37	[5]
	Drip	2740	500	4450	4539	2,19,200	3.42		
Pea	Conventional	58,700	400	4000	4080	46,96,000	81.5	24.00	[16]
	Drip	77,400	400	3040	3100.8	61,92,000	141.4		

through drip irrigation is the best choice. More is the energy utilization for lifting of water, more would be operational cost, such as for diesel or electricity.

Figure 15.1 shows that water-energy productivity was highest under drip irrigation system for all vegetable crops. Scientists[7,19] have reported that energy input equivalent for irrigation (m³) is about 1.02 MJ/ha compared with 0.80 MJ/ha of energy output for vegetable production.[7,19] The total energy inputs (MJ/ha) for conventional and drip irrigation methods were:

- Capsicum: 7650 and 3051;
- Chili: 5100 and 4539;
- Cucumber: 5100 and 4233;
- Okra: 8160 and 4406;
- Pea: 4080 and 3100;
- Potato: 7140 and 4569; and
- Tomato: 6120 and 3243.

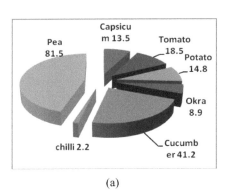 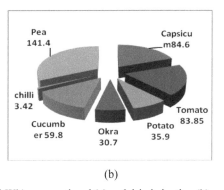

(a)   (b)

FIGURE 15.1   Water energy productivity (kg/kWh): conventional (a) and drip irrigation (b).

The total energy outputs (MJ/ha) for conventional and drip irrigation were estimated as:

- Capsicum: 1,456,000 and 3,648,000;
- Chili: 159,200 and 219,200;
- Cucumber: 2,968,000 and 3,576,000;
- Okra: 1,028,800 and 1,912,800;
- Pea: 4,696,000 and 6,192,000;
- Potato: 1,488,000 and 2,316,000; and
- Tomato: 1,600,000 and 3,840,000.

15.4   FERTILIZER USE EFFICIENCY

Appropriate selection of fertilizers and timing of their application are essential to harness higher crop yield under drip irrigation (Table 15.2). The water solubility, compatibility with other fertilizers, their chemical reaction with water, water temperature, etc. are essential characteristics that must be considered while selecting inorganic fertilizers for fertigation. The drip fertigation enhances the root activity and improves the mobility of nutrients resulting in enhanced nutrient uptake. Hundred percent water soluble fertilizers generally safe guard the drip system from clogging in the long run.[6]

TABLE 15.2   Fertilizer Use Efficiency (FUE) under Conventional and Drip Irrigation Methods.

Crop	Irrigation method	Crop yield (kg/ha)	Total quantity of nutrient applied (kg/ha)			FUE (%)
			N	P	K	
Capsicum	Conventional	18,200	250	150	150	33.09
	Drip irrigation	45,600				82.91
Tomato	Conventional	20,000	250	150	250	30.76
	Drip irrigation	48,000				103.08
Potato	Conventional	18,600	120	240	120	38.75
	Drip irrigation	28,950				60.31
Okra	Conventional	12,860	200	100	100	32.15
	Drip irrigation	23,910				59.78
Cucumber	Conventional	37,100	150	75	75	123.67
	Drip irrigation	44,700				149.00
Chili	Conventional	1990	120	80	80	6.78
	Drip irrigation	2740				9.78
Pea	Conventional	58,700	60	80	70	279.52
	Drip irrigation	77,400				368.57

The properly designed drip fertigation system should deliver water and nutrient at a rate, duration, and frequency to maximize crop water and nutrient uptake, while minimizing leaching of nutrients and chemicals. Application of NPK with liquid fertilizer can significantly increase the concentration and total uptake of N, P, and K in vegetable crops compared with solid fertilizers.

Research study[18] indicated that drip fertigation resulted in approximate 40% saving of water (Figs. 15.2 and 15.3).

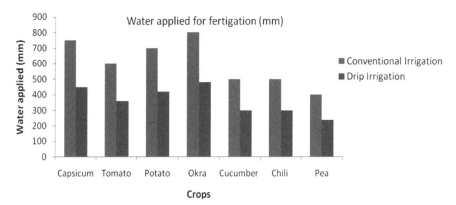

FIGURE 15.2   Water applied under drip fertigation compared with conventional irrigation.

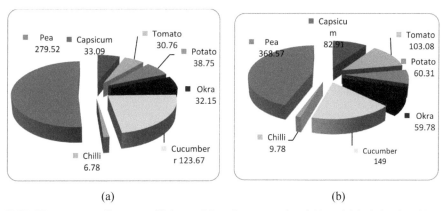

(a)   (b)

FIGURE 15.3   Fertilizer use efficiency (%) under conventional (a) and drip irrigation (b).

15.5   LABOR USE EFFICIENCY

Selection of any irrigation system depends mostly on its efficiency, labor requirement, and initial investment apart from other savings. The initial investment of drip irrigation systems is high. However, due to efficacy in minimizing water and nutrient losses with reduced labor, the Government of India is promoting adoption of drip irrigation systems for horticultural and agricultural crops.

The conventional irrigation practices warrant labor requirement in every season and every day while irrigating the fields. In order to apply water in conventional irrigation system, land grading and formation of furrows/basins are essentially time- and labor-consuming activities. Several studies have proven the advantages of labor saving with drip irrigation systems compared with conventional irrigation method (Table 15.3). In capsicum (25%), tomato (25%), potato (19%), okra (13.8%), cucumber (15.7%), chili (21.8%), and pea (19.23%), savings in labor cost are possible in drip irrigation compared with conventional irrigation.

TABLE 15.3 Labor Requirements (Man-Days) under Conventional and Drip Irrigation Practices.

Vegetable crop	Labor requirements, man-days		Labor saving over conventional (%)	Ref.
	Conventional irrigation	Drip irrigation		
Capsicum	185	160	25	[12]
Chili	128	100	21.8	[8]
Cucumber	121	102	15.7	[3]
Okra	185	160	13.8	[2]
Pea	130	105	19.23	[3]
Potato	124	105	19	[10]
Tomato	194	169	25	[12]

15.6 SUMMARY

With present-day practices of flood irrigation systems with 25–40% efficiency, it would be difficult to assure the food security for the ever increasing population under scarcity of natural resources. Adoption of micro irrigation systems is the viable solution to address the depletion of water resources, due to significant saving of water/fertilizers/labor/energy and superior quality and quantity of produce to achieve "more crop per drop." Therefore, wider adoption of this technology among the farming community in India would help in maximizing the input (water, fertilizer, labor, and energy) use efficiency.

KEYWORDS

- **drip irrigation**
- **energy use efficiency**
- **fertilizer use efficiency**
- **labor use efficiency**
- **water use efficiency**

REFERENCES

1. Bharadwaj, R. L. Effect of Mulching on Crop Production under Rainfed Condition-A Review. *Agric. Rev.* **2013**, *34*, 188–197.
2. Esmat, A. B.; Nozrul Islam, M. D.; Alam, Q. M.; Byazid, H. S. M. Profitability of Some Bari Released Crop Varieties in Some Locations of Bangladesh. *Eur. J. Agron.* **2011**, *36* (1), 111–122.
3. Everaarts, H.; Maerere, A. P. *Profitability, Labor Input, Fertilizer Application and Crop Protection in Vegetable Production in the Arusha Region, Tanzania.* Wageningen, 2015; pp 1–40.
4. Fathia, E. M.; Kamel, N.; Mohamed, M. M.; Netij, B. M. Effects of Surface and Subsurface Drip Irrigation Regimes with Saline Water on Yield and Water Use Efficiency of Potato in Arid Conditions. *Tunisia J. Agric. Environ. Int. Dev.* **2014**, *108* (2), 227–246.
5. Gulshan, M.; Singh, K. G.; Sharda, R; Siag, M. Response of Red Hot Pepper to Water and Nitrogen under Drip and Check Basin Method of Irrigation. *Asian J. Plant Sci.* **2007**, *6* (5), 815–820.
6. Hebbar, S. S.; Ramachandrappa, B.; Nanjappa, H. V.; Prabhakar, M. Studies on NPK Drip Fertigation in Field Grown Tomato. *Eur. J. Agron.* **2004**, *21*, 117–127.
7. Ibrahim, H. Y. Energy Use Pattern in Vegetable Production under Fadama in North Central Nigeria. *Trop. Subtrop. Agroecosyst.* **2011**, *14*, 1019–1024.
8. Jagtap, P. P.; Wagh, K. K.; Shingane, U. S.; Kulkarni, K. P. Economics of Chilli Production in India. *Afr. J. Basic Appl. Sci.* **2012**, *24* (5), 161–164.
9. Jat, R. A.; Wani, S. P.; Sahrawat, K. L.; Singh, P.; Dhaka, B. L. Fertigation in Vegetable Crops for Higher Productivity and Resource Use Efficiency. *Indian J. Fert.* **2011**, *7*, 22–37.
10. Kaur, J.; Quadri, J.; Ahmad, P.; Nafees, A.; Chesti, M. H.; Ahmad, H. S.; Bhat, B. A. Study on Economics of Potato Growing Towards Livelihood Security. *Afr. J. Agric. Res.* **2013**, *8* (45), 5639–5644.
11. Khan, S.; Khan, M. A.; Hanjra, M. A.; Mu, J. Pathways to Reduce the Environmental Footprints of Water and Energy Inputs in Food Production. *Food Policy* **2009**, *34* (2), 141–149.

12. Khatri, R. T.; Mistry, H.; Patel, K. S. Comparative Economics of Production of Important Vegetables in Surat District International Research. *J. Agric. Econ. Stat.* **2011**, *2* (1), 58–62.

13. Makau, J. M.; Masinde, P. W.; Home, G. P.; Njoroge, C. K.; Mugai, E. N. Growth and Yield of Tomato under Alternate Furrow Irrigation. *Afr. J. Hortic. Sci.* **2014**, *8(2)*, 24-37.

14. Muhammad, A.; Malik M. A.; Rana, M. A.; Muhammad, A. R. Impact of Drip and Furrow Irrigation Methods on Yield, Water Productivity and Fertilizer Use Efficiency of Sweet Pepper (*Capsicum annuum* L.) Grown under Plastic Tunnel. *Sci. Lett.* **2016**, *4* (2), 118–123.

15. Nadiya, N.; Kurien, E. K.; Mathew, E. K.; Varughese, A. Impact of Fertigation and Drip System Layout in Performance of Chilli (*Capsicum Annuum* L.). *Int. J. Eng. Res. Dev.* **2013**, *7*, 85–88.

16. Narayanamoorthy, A.; Deshpande, R. S. Irrigation Development and Agricultural Wages: An Analysis across States. *Econ. Pol. Wkly.* **2003**, *38* (35), 3716–3722.

17. Patel, J. C.; Patel, B. K. Response of Potato to Nitrogen under Drip and Furrow Methods of Irrigation. *J. Indian Potato Assoc.* **2001**, *28* (2–4), 293–295.

18. Patel, R. S.; Patel, P. G. Effect of Irrigation and Nitrogen on Growth and Yield of Brinjal under Drip System. *J. Maharashtra Agric. Univ.* **2006**, *31*, 173–175.

19. Rafiee, S.; Mousavi Avval, S. H.; Mohammadi, A. Modeling and Sensitivity Analysis of Energy Inputs for Apple Production in Iran. *Energy* **2010**, *35*, 3301–3306.

20. Stewart, W. M.; Reiter, J. S.; Krieg, D. R. Cotton Response to Multiple Applications of Phosphorus Fertilizer. *Better Crops* **2005**, *89*, 18–20.

21. Tiwari, K. N.; Singh, A.; Singh, A.; Mal, P. K. Effect of Drip Irrigation on Yield of Cabbage under Mulch and Non-mulch Conditions. *Agric. Water Manage.* **2003**, *58*, 19–28.

RAINWATER CONSERVATION AND UTILIZATION TECHNIQUES: THE NORTH-EAST HILLY REGION IN INDIA

JOTISH NONGTHOMBAM and SANTOSH KUMAR

ABSTRACT

North East Hilly (NEH) Region of India is characterized by heavy rainfall and varying topography. Investment in terms of micro irrigation kits for MWH (micro water harvesting) was around Rs. 1050.00, with high-density polyethylene lining cost was Rs. 6608.00 per 16 m^3/s of capacity. Vegetables under irrigation systems during nonrainy reason showed encouraging results. A micro-level water conservation model and utilization technique under tough hilly terrain of Mizoram, NEH Region in India is discussed in this chapter.

16.1 INTRODUCTION

The adverse effects of changing climate, growing population has imposed immense pressure on advances in agricultural practices. The importance of micro-level approach in developing techniques to support income generation for regional farming communities is imperative.[11,12]

In the course of developing an effective farming system that can counter multiple problems of changing climate, food security, sustainability, etc., researchers and scientists around the globe have recommended Integrated Farming System (IFS) approaches. Transition to such new farming system from traditional farming in an open environment can greatly increase the possibility of achieving food security, sustainable, and climatic resilient agriculture. Keeping all these subtleties, integration of coupled micro water harvesting (MWH) and rooftop rain water harvesting (RRWH) systems in evolving IFS is vital and essential.

Approximately existing 2.5 million traditional ponds in India (either manmade or natural) are generally being adopted for domestic, agriculture, and livestock purposes.[2,5,8] North East Hilly (NEH) Region of India is characterized by varying topography that is largely affected by high seepage and surface flows. Moreover, dual effects of water in the form of heavy rainfall during monsoon and water scarcity during post-monsoon are severe in this region. Existing undulated terrains and dual effects of water are the main limiting constrains in storing, conservation of rain and runoff water for its later use in domestic, agriculture, and livestock enterprises.

Nongthombam et al.[6] presented a detailed discussion on the potential of rain water harvesting and its later use in NEH, India from settled agriculture in the form of irrigated terrace in parts of Nagaland and Meghalaya to bamboo irrigation system found in parts of Meghalaya and in some villages in Mokokchung district of Nagaland. A trend of good stride with respect to RRWH is found in Mizoram and Sikkim. Although traditional methods are supporting the indigenous farming systems, present advances in water resources demand for supporting domestic consumption, agriculture, and livestock requirements, yet the scientific approach to water conservation techniques has become a necessity in NEH region of India.

Traditional water harvesting methods adopted are exposed to potential losses such as leakage in storage tanks, faulty designs, infiltrations, percolation, seepage flows, and evaporation. Also high initial monetary requirement for implementing MWH, RRWH using concrete pond, tanks makes it difficult to adopt for the farming community of NEH region in India. Therefore, it is essential to propagate micro-level approach to economic base coupled MWH and RRWH systems in the process of transition from traditional farming system to sustainable farming system. With an effort to review the existing MWH/RRWH and the technical aspects involved in it, an overview of recommended techniques is presented in this chapter.

This chapter reviews design aspects of the existing MWH and RRWH techniques suitable for NEH regions in India. Advantages of micro-level water conservation model and utilization techniques under tough hilly terrain of Mizoram, NEH Region in India are also discussed. The model presented in this chapter can be effective in uplifting the livelihood of farming communities of this region.

16.2 ROOF-TOP RAIN WATER HARVESTING

Collection of rain drops that fall on the roof of any premises/building and further storage/concentration into a storage tanks, pond, reservoirs by means of conveyances with different shaped gutters and pipes is termed as RRWH. NEH region of India is characterized by the dual effects of water and geological formations. Therefore, RRWH techniques prove vital and essential to tackle severe water scarcity during off-rainy season. The advantages and limitations of rooftop water harvesting are[1,10]:

- Contributes to meet the increasing water demand;
- Recharges ground water resources;
- Reduces runoff and prevents chocking of drains;
- Provides self-sufficient for water availability;
- Reduces potential of soil erosion;
- Being adjacent to the house/premises, water is easily available for use;
- Being personal ownership, the maintenance of the system can be done in better way;
- The construction cost is lower than the water collection by motor, pumps, etc.;
- The materials required for construction are easily available at reasonable price/cost.

On the other hand, following bottlenecks also adhere to this system:

- Pollutants in air gets mixed in the rainwater that can lead to health problems. India cities such as Ludhiana (Punjab), Kanpur (U.P.) are facing such problem.
- Water quality deterioration by (1) increasing level of nitrates and fluorides in the water; (2) chances of high contamination of cadmium, lead, iron, and chromium in the stored water; (3) increase in bacteriological contamination levels in the stored water; (4) other main reasons are irregularities in intervals/variation of the rainy season; (5) uneven distribution of rainfall makes design task of RRWH very tedious, if it is to conserve water for long period of time.

16.2.1 AMOUNT OF WATER TO BE HARVESTED

The estimation of water to be harvested in RRWH system is essential to have an idea of the capacity of the water that can be harvested annually.

The volume/amount of water to be harvested may be estimated using the following equation:

$$\text{Total water harvested (L/year)} = \left[A \times R \times C_R \times C_F \right] \times 1000 \qquad (16.1)$$

where A is the total guttered roof area (m^2); R is the total annual rainfall (m); C_R is the runoff coefficient; C_F is the filtering coefficient (Generally taken as 0.9); 1000 is the conversion factor to convert m^3 into liters.

Basically, the runoff coefficient considers the evaporation losses varying from 0 to 1.0.[11] For different roof surfaces, the runoff coefficients (C_R) are given in Table 16.1.

TABLE 16.1   Runoff Coefficient Values.

Roof surface	Runoff coefficient (C_R)
Asbestos sheet	0.8–0.9
Galvanized iron sheet	>0.9
Organic material: Thatch, Palm, etc.	0.2
Tile	0.6–0.9

Source: Ref. [11].

16.2.2   COMPONENTS OF RRWH

16.2.2.1   ROOF SURFACE

A roof should be suitable to yield water. Roofs made of tiles, metal sheets, and asbestos sheets are most suitable. Generally the cumulative rainwater of a roof in a year may be calculated as the product of annual rainfall and the roof area.

16.2.2.2   GUTTER

Gutter of different shapes is used to carry the harvested water from roof to the storage tank. Gutters are generally available in semicircular or trapezoidal, U or V shapes either made of metal sheets or plastic cast.[11] Depending on the size of the of the gutter, corresponding down-pipe size are given in Table 16.2.

TABLE 16.2   Recommended Gutter and Down-Pipe Size.

Roof area (m²)	10	13	17	21	29	34	40	46	66
Trapezoidal gutter (mm)	50	55	60	65	75	80	85	90	100
Down-pipe size (mm)	15	20	25	25	32	32	40	40	40

Source: Ref. [11].

16.2.2.3   FILTER

To avoid collection of waste materials and contamination of pollutants, desired filters can be installed at the junction of the gutter and outlet of the down-pipe, respectively.

16.2.2.4   STORAGE TANK FOR WATER

Harvested water from the roof area can be stored in storage tanks, such as molded plastic, drum, open-frame ferro-cement, plate, interlocking block, and brick-lime cistern tanks. Harvested water adopting rooftop harvesting techniques have great potential in NEH region in India, where existing undulated terrain and dual effects of water in the form of flash flood and severe drought are main limiting constrains in storing, concentrating rain water for later use in domestic consumption, agricultural, and livestock purposes. Further, RRWH technique coupled with micro water harvesting system, *Jalkund*, can greatly enhance the scope for adoption of micro-level approach to develop sustainable and climatic resilient IFS.

16.3   MICRO WATER HARVESTING

The farm ponds for water harvesting with different and multiple purposes are generally adopted/constructed to serve as recreation pond, domestic and livestock purposes, or fish production. It can be broadly classified as embankment type, excavated, or dug-out type.[4]

16.3.1   EMBANKMENT POND

A water body established by means of construction of embankment across stream/water path can be termed as an embankment pond.[1] Where topographical gradient ranges from gentle to moderate steep with sufficient depression in stream valleys, such ponds are recommended.[4,10]

16.3.2   EXCAVATED POND

A water body developed in the form of pit dug-out excavating the designed earthwork can be termed as excavated pond. Generally such ponds are recommended for relatively plain surface terrains.[4,10]

16.3.3   POND DESIGN

A well-designed farm pond is an important component in establishing an IFS that will be socioeconomically viable for the indigenous farming community of the region. Parameters—such as the geomorphological characteristics of the catchment area, rainfall–runoff relationship, water requirement, and losses (evaporation/infiltration/percolation/seepage)—play key role in understanding and working out of a thoughtful design for a farm pond.[1,4,10] Following basic steps can be followed in designing a farm pond.

16.3.3.1   LOCATION

Pond site selection should be initiated by surveying and studying the area having at least more than one site. Thereafter individual sites should the ranked in terms of most feasible and economical when implemented.[1,3] Obtaining larger storage volume capacity with minimal earthwork is essential and such conditions can be found in narrow valleys with relative steep slopes. Further, such sites should be judiciously studied for any unfavorable geological conditions.[1]

- Irrigation farm ponds should be developed relatively near the cultivation areas under irrigation. If fishery component is to be included in the pond design, then it has to be readily accessible for transportation.
- Site selection should be such that farm sheds, sewage, and industrial waste drainages will not pollute the farm pond. However, if unavoidable, then drains are needed to be diverted from the design farm pond to avoid possible contamination. Cables and pipe lines beneath the design farm ponds must be avoided.
- The location of the pond should be such so that sudden failure of the embankment would not harm and cause loss to life, damage to properties or hindrance to public services.
- Bunding of the periphery of the pond can be done by the excavated soil. Suitable inlets, spillways, and other essential structures should be included in the design.
- Acceptable similarity in the volume of excavated soil and fill soil should be considered while selecting the site.

16.3.3.2   SEEPAGE LOSSES

If the designed pond is surface flow faded, then the site should be selected, where soil of impervious characteristics is available to mitigate possible excess seepage losses. Farm ponds subjected to high seepage losses may be designed with cement, polyethylene, brick, slabs linings.[1,3] Farm ponds in steep hilly terrains may also be lined with suitable polyethylene beneath and brick or boulders above it to reduce wear tear and seepage.

16.3.4   CAPACITY OF A POND

Based on the catchment size and area to be covered, the capacity is estimated for surface runoff farm ponds.[1,4] For irrigating farm ponds or runoff ponds, the same is designed on the basis of the irrigation demand of the cultivation area.[1,4] Generally for trapezoidal or rectangular pond, the capacity (v) can be determined using the following equations[1,3,4]:

- For trapezoidal pond with side slope 1:1

$$v = \frac{h}{2}\left[(a-2f)(1-2f)+(1-2h-2f)(a-2h-2f)\right] \qquad (16.2)$$

- For rectangular pond

$$v = \frac{h}{2}\left[la+(1-2h)(a-2h)\right] \qquad (16.3)$$

where l is the top length (m); a is the top width (m); h is the pond depth (m); and f is the free board (m).

The minimal depth of the farm pond should be 1.5 m to avoid excessive water spread area with upper height not exceeding 5.0 m.

16.3.5   PLASTIC MATERIAL FOR LINING

Materials such as polyethylene or polyvinyl chloride (PVC) are widely used for lining of pond. The desirable water barrier characteristics are:

- degree of seepage control that can be expected;
- soil microorganisms deterioration resilient, atmospheric elements, and movement of subgrades;
- wear and tear;

- toxicity;
- lesser complexity in installation;
- easy in transportability;
- economical.

16.3.6   REQUIRED MATERIAL SIZE

The required material dimension/size of the pond can be estimated based on Bureau of Indian Standards 2009-IS-15828 using the following equation:

$$\text{Film area} = [2(2^{0.5})\,(h+f)+(a-2\,h-2f)]$$
$$\times\ [2(2^{0.5})\,(h+f)+1+(1-2\,h-2f)] \tag{16.4}$$

where l is the top length (m); a is the top width (m); h is the depth of the pond (m); f is the free board (m).

16.3.7   JOINING OF THE FILM LINING: METHODS

It is always best if the film lining is procured in whole to avoid joining separate films. In case if the required lining film size is not available as a whole, then the required size can be obtained by joining separate films together. Methods for joining the lining film are described in this section.

16.3.7.1   HEAT SEALING OR THERMAL WELDING

Using one or two heated blades adjusted at suitable temperature (Fig. 16.1), pressure is applied to the prespecified portion of the lining film.[1] With this method, polytetrafluorethylene impregnated, coated glass-cloth must be placed between the adjacent lining films to protect it from getting spoiled, if it sticks to the blades. Usually this method is a most commonly adopted technique.

16.3.7.2   HOT BITUMEN JOINING

This method uses bitumen coat (grades of 85/25 and 80/100 in a ratio 2:1, respectively) heated at 100°C for joining separate lining films.[1,3] Prior to application of heated bitumen coat, it is first tested on a small piece of film to avoid damaging of the film thereafter on 10-cm area along the width of the film and then is folded (Fig. 16.2) and is covered with bricks to avoid direct exposure.[1,3]

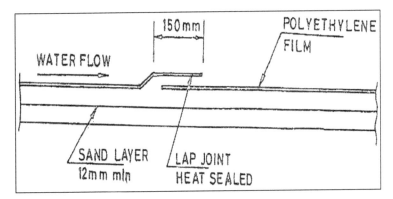

FIGURE 16.1   Heat sealing (thermal welding).

Source: Reprinted from Ref. [1].

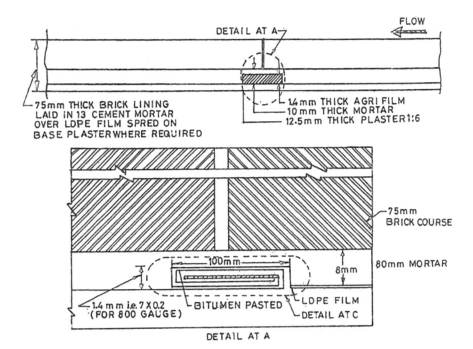

FIGURE 16.2   Hot bitumen joining.

Source: Reprinted from Ref. [1].

16.3.7.3   OVERLAPPING

Overlapping of the separates lining sheets (Fig. 16.3) is permitted to avoid leakage of water. In doing so, the overlapping section should not be less than 30 cm for earth cover and 15 cm for hard cover. Generally this method is least recommended as it is prone to leakage problems.

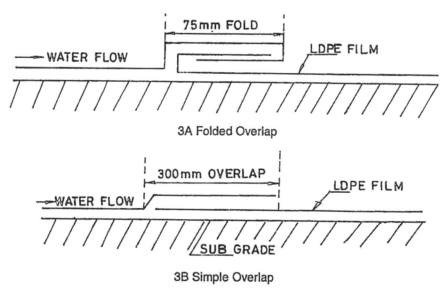

FIGURE 16.3   Overlapping method.

Source: Reprinted from Ref. [1].

16.3.7.4   OVERLAP JOINTS WITH PRESSURE TAPE

The adjacent separates lining films can be joined using pressure tapes (Fig. 16.4). Generally this method is simple, if the pressure tape is easily available locally.[1]

16.3.8   POND SEALING OR LINING OR FLEXIBLE MEMBRANE

Farm ponds or irrigation ponds can be protected[7] using lining materials (Table 16.3) for restricting, controlling of water losses from deep percolation and seepage, pollution from waste contamination structures water conservation.

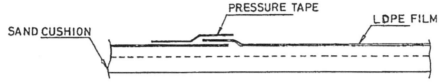

FIGURE 16.4 Over-lap joint with pressure tape.

TABLE 16.3 Recommended Pond Liners.

Type	Detail	Minimum thickness criteria	
		Waste water	**Clear water**
EPDM	Ethylene propylene diene: Terpolymer geomembrane	45 mil	45 mil
FPP	Flexible polypropylene: Geomembrane	40 mil	30 mil
FPP-R	Flexible polypropylene: Reinforced geomembrane	36 mil	24 mil
HDPE	High-density polyethylene: Geomembrane	40 mil	30 mil
LLDPE	Linear low-density polyethylene: Geomembrane	40 mil	30 mil
LLDPE (R)	Linear low-density polyethylene: Geomembrane (reinforced)	36 mil	24 mil
PVC	Polyvinyl chloride: Geomembrane	40 mil	30 mil

1 mil = (1/1000) in.

Source: Ref. [9].

16.3.9   EARTH COVER

- To avoid any damage to the bottom portion of the sheet layer, sieved/excavated soil can be spread to a depth of 5 cm and later compacted over the film.
- A 15-cm layer of the excavated soil then can be covered and rammed after being soaked with water.
- The side-wall slopes then can be turfed with locally available *durba*, dry grass, dried pine leaves, thatch grass, etc.

16.3.10   PLASTIC FILM LINING WITH STONE PITCHING

Plastic lining with stone pitching can be accomplished by certain steps of subgrade preparation, covering through film and finally pitching of suitable stones.[1,3]

16.3.11   SUBGRADE PREPARATION

- The excavated pond should be at least made 60-cm deeper than the designed bed level.
- Perimeter soil should then be properly compacted and also should not have any sharp edges to avoid tear or puncture of the lining sheet.

16.3.12   COVERING THROUGH THE LINING FILM

- A layer of fine sand of 12-mm thickness can be laid on the top of subgraded layer to avoid any damage to the lining film.
- The lining film can then be spread. The outer end of the film should be anchored properly with weights being placed at every finished lining until the whole lining process is finished. When the laying of the film is complete, it should be sufficiently loose enough leaving around 5% extra length of the lining film.

16.3.13   STONE PITCHING

The side wall of the lined pond can be covered by smooth riverside stones, boulders, flat stones, or bricks.[1] In doing so, any possible sharp edges need to be chipped off and need to be smoothen. In the process of stone pitching, usually thicker stones of size 20 cm should be placed at the bottom of the lined pond and then proceeding upwards, and the stones should be of smaller size so that the stones at the top of the pond periphery are of 10-cm sizes. When the stone pitching is done, it should provide firm grip to each other. The existing voids between the stones are then filled with the excavated soils to give more stability. Finally at the top periphery of the pond, preferably 30-cm wide stone-pitching is done with a cement mortar in a ratio of 1:6.[1,3]

16.4   POND MANAGEMENT

Proper inspection and maintenance in regular intervals is essential for prolonged life span of the farm pond. Such observations and maintenance are included in the inspection of the pond periphery, finding problems and rectifying. Some of the important considerations are:

- The surrounding area of the farm pond should be protected from wild animals by providing fencing using locally available material such as bamboo, barbed wire, tree trunk;
- Vigorously growing trees must always be removed or avoided in the surrounding of the pond to avoid damage of the pond;
- Side filling and any existing slope of the farm pond can be protected from possible erosion by covering with suitable fodder or any other normal grass or vegetative cover;
- If aquatic vegetation is added in the pond, then periodical reduction of its population is essential;
- Removal of soil sediments from the bottom of the pond can be done annually to avoid reduction in storage capacity and possible wear tears.

16.5   WATER HARVESTING SYSTEMS: ROOF TOP (RRWH) VERSUS MICRO (MWH)

Implementation, adoption of roof top (RRWH) and MWH techniques in India can be found in wide ranges. Researchers and scientists have been adopting RRWH and MWH systems to promote and support different farming systems under a wide variation of topography and climatic stress.

With an effort to highlight the applicability of RRWH and MWH systems in the process of transition from traditional farming system to sustainable and climatic resilient IFS, some of the existing techniques suitable in India for farming communities of NEH Region in India are (1) at Dimapur, Nagaland, an economical integrated ways of farming system coupling rainwater harvesting technique; (2) Jalkund and micro irrigation system using gravity drip/sprinkler systems to support duck base IFS for 30 farming families in Phek District of Nagaland in India.

The water harvesting pond (Jalkund of cross-sectional dimensions of 5 m × 4 m × 1.5 mwith side slope 1.5:1) was constructed in the farm with a thickness cushioning of 3–5 using dry pine leaf or thatch grass and then was lined with low-density polyethylene (LDPE) 250-μm liner. Figure 16.5 shows smoothing, cushioning, and lining of Jalkund.[4]

An integration of RRWH and MWH systems to support cultivation under polyhouse system in mid-hills of Uttarakhand was adopted. Harvested water from the roof top of 200-m^2 polyhouse was stored in a water harvesting pond with total effective capacity of 37,000 L. Figure 16.6 shows the hybrid capsicum under polyhouse supported with MWH system.[10]

FIGURE 16.5 Smoothing, cushioning, and lining of Jalkund.
Source: Reprinted with permission from Ref. [4].

FIGURE 16.6 Hybrid capsicum under polyhouse system supported with MWH system.
Source: Reprinted with permission from Ref. [10].

A detailed implementation strategy and application of gravity-type drip irrigation system (GDS) and its integration with Jalkund at the farm under tough hilly terrains of Aizawl district in Mizoram, India is shown in Figure 16.7.[6]

FIGURE 16.7 Gravity-type drip irrigation system (GDS) with MWH technique, Jalkund.
Source: Reprinted with permission from Ref. [10].

The State Directorate of Water Management presented an extensive report on both RRWH and MWH systems, their components and design for southwestern part of Punjab, India. They also presented MWH lined with LDPE/high-density polyethylene (HDPE) covered with stone and boulder pitching.[9] Figure 16.8 shows water harvesting farm ponds LDPE lined with stone and boulder pitching.

FIGURE 16.8 Water harvesting farm LDPE ponds lined with stone and boulder pitching.
Source: Reprinted from Ref. [3].

A wide range of different capacity water harvesting ponds[3] for 78 farming families for cultivating high-value vegetable crops under open and protected cultivation systems in villages of Nanital and Almora district are shown in Figure 16.9.

FIGURE 16.9 Water harvesting farm ponds with low cost polyhouse system.
Source: Reprinted from Ref. [3].

16.5.1 APPLICATIONS

MWH ponds lined with HDPE integrated with gravity drip and mini sprinkler systems were implemented at 50 farms in Aizawl district in Mizoram

(Figs. 16.10 and 16.11). The region is prone to intense dual effects of water (flood and drought during monsoon and off-monsoon season, respectively) subjected with tough hilly terrains.

Therefore, it is important to identify and propagate a suitable low cost technique that can be easily implemented and replicable for the indigenous farming communities. Attempt was made to propagate the concept of replicable economical water conservation and utilization techniques for enhancing the farming scenario of the NRE region.

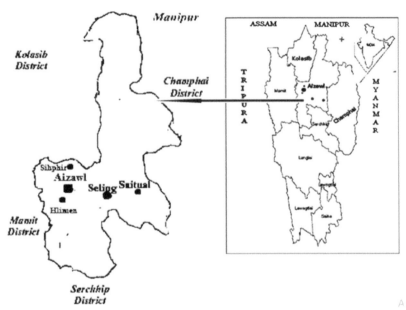

FIGURE 16.10   Aizawl district map in Mizoram.

FIGURE 16.11   Water harvesting pond at the farm.

The region mainly depends on rain-fed traditional terrace farming system, where the terraces are only 1–1.25 m. During off-monsoon season due to severe scarcity of irrigation water, agricultural production declined drastically. With suitable micro water harvesting technique and simplified gravity micro irrigation technique, the effects of dual impact of water to agricultural practices can be mitigated. Under participatory approach, 50 Jalkunds (dimension 5 m × 4 m × 1.5 m with capacity of 27,000 L lined with HDPE-5 layers of 300-μm geomembrane) were constructed to harvest water (Fig. 16.13).

FIGURE 16.12 Preparation of outlet and stands for storage tank at the farm.

Further, micro gravity drip and sprinkler system on 30 and 90 m², respectively were also integrated with the harvesting pond to promote irrigation system. The irrigation systems were designed with proper joints (laterals and T), so that the unit can easily be dismantled and sifted to another terraces. For implementing the GDS, a layout of the design drip system was worked out at the farm with specific plant-to-plant and row-to-row spacing. Storage tank height of 1–1.5 m was estimated depending on the size and pattern of the layout and outlet for main line is prepared (Fig. 16.12). Using fittings, T and couple fittings were connected with the tank outlet, main line, and lateral lines (Fig. 16.13).

FIGURE 16.13 Tank outlet, main, and lateral lines connected using fittings.

Using locally available bamboo, stakes and clips are prepared. These stakes and clips are used to place the pipeline (main and lateral lines) in place according to the layout (Fig. 16.14). Similarly for sprinkler system, the layout was prepared at farm (Fig. 16.15). Coupling micro water harvesting technique, *Jalkund* with gravity drip and mini sprinkler irrigation system can greatly uplift the scope of adoption of micro irrigation system, which in-turn can enhance the cultivation practices during off-season (Rabi: November–February) in these regions with lesser complexity. Tables 16.4–16.6 show construction cost of micro water harvesting with gravity drip and mini sprinkler systems.

FIGURE 16.14   Staking and clipping of the lateral lines of GDS.

FIGURE 16.15   Gravity-based mini sprinkler (blue and black) system at farmers' farm.

TABLE 16.4   Construction Cost of *Jalkund* System.

Particulars of a pond	Rate (Rs.)	Amount (Rs.)
Volume of excavation: 16 m³/s	Farmer's participation and KVK assistance	
Planking and plastering the sides and bottom		
Cost of HDPE geomembrane (300 μm):size (8 × 7 m²) = 56 m².	118.00/m²	6608.00
Total per unit	–	**6608.00**

TABLE 16.5 Construction Cost of Gravity Drip System (GDS).

Particulars	Rate (Rs.)	Amount (Rs.)
Gravity drip irrigation, GDS: area of 30 m²	1300.00	1300.00
Quoton (100 L)	1100.00	1100.00
Terracing	Farmer's participation and KVK	
Wood lock for erecting the water tank: (1.5 m)	assistance	
Laying out and installation		
Total (Rs. per unit)		**2400.00**

TABLE 16.6 Construction Cost of Micro Sprinkler System (GDS).

Items with units	Quantity	Rate (Rs.)	Amount (Rs.)
Mini sprinklers with stalks	10	38.00 each	380.00
Tee, 16 mm	10	07.00 each	70.00
Lateral pipe, 16 mm	40 m	15.00/m	600.00
Total (Rs. per unit)			**1050.00**

16.5.2 CLIMATIC CONDITIONS[13]

The meteorological data prevailing during 2015–2016 at the study sites are presented in Table 16.7.

TABLE 16.7 Climatic Data during the Cropping Period (2015–2016) at Sihphir, Aizawl, Mizoram Sites.

Month	Temperature (°C)		R.H. (%)	Rainfall (mm)
	Minimum	Maximum		
April 2015	13.9	32.3	81.98	374.8
May 2015	15.7	31.7	70.32	245.1
June 2015	17.8	30.6	75.48	412.0
July 2015	18.7	30	93.05	504.3
August 2015	18.5	30.2	93.35	589.1
September 2015	18.5	30.5	93.02	303.1
October 2015	16.9	30.4	91.56	219.6
November 2015	13.5	28.9	90.03	4.2
December 2015	10.2	25.7	90.42	0.8
January 2016	26.91	11.50	67.11	0.0
February 2016	26.79	13.02	55.75	25
March 2016	32.38	17.10	61.20	37

16.5.3 SOIL PROPERTIES

To determine the initial fertility status, a composite soil sample was taken just after the field preparation and before laying out the experiment with the help of soil auger from the field at soil depth of 0–25 cm. The physiochemical properties are given in Table 16.8.

TABLE 16.8 Physicochemical Properties of Soil at Sihphir, Aizawl, Mizoram Sites.

Soil property	Value
Sand percent	61.99
Silt percent	23.20
Clay percent	14.81
Texture class	Sandy loam
pH	5.9
Electrical Conductivity	0.31
Organic carbon percent	0.60
Nitrogen (kg/ha)	371.40
Phosphorus (kg/ha)	102.20
Potassium (kg/ha)	319.80
Sulfur (mg/kg)	14.90
Zinc (mg/kg)	7.20
Boron (mg/kg)	2.30
Iron (mg/kg)	24.60

16.6 DISCUSSIONS

The presented model was implemented at 50 farms in Dakla zau, Buh ching pawl, and Tam lowng veng farming communities of Sihphir venghlun, Aizawl district in Mizoram. With HDPE-lined harvesting pond, rain and runoff water are easily harvested and used for irrigation during off-season. Further stored water is effectively used during off-season for irrigating the *Rabi* crops, which can uplift the scope of agricultural production of farmers. Individual details of farm record book are being maintained by the farmer under the program to have the insight of income under the proposed model. Table 16.9 shows the details of income generated from cultivation of vegetable crops throughout the year for 10 farmers within the group to show an insight of the income (2015–2016).

TABLE 16.9   Income Generated by Cultivation of Vegetable Crops throughout the Year for 10 Farmers.

Farmer	Average area (m2)	Crops cultivated	Quantity (kg)	Rate (Rs./kg)	Income (Rs. per unit)
Vanlalthlanga	500	Chow chow	150	15	2250
	500	Broccoli	500	100	50,000
H. Zirsangliana	250	Chow chow	60	15	900
	300	Broccoli	250	100	25,000
	350	Leafy mustard	1000	35	35,000
	500	French bean	350	50	17,500
S. Vanzinga	500	Chow chow	150	15	2250
	700	Leafy mustard	2000	40	80,000
	500	Bean	400	50	20,000
	250	Coriander	500	100	50,000
Rochhungi	350	Leafy mustard	1000	40	40,000
	750	Bean	500	50	25,000
	300	Coriander	600	100	60,000
	100	Bitter gourd	200	30	6000
K Lalchamliana	150	Leafy mustard	500	40	20,000
	250	Bean	200	50	10,000
	150	Coriander	300	100	30,000
	50	Gourd	100	30	3000
	150	Cabbage	250	30	7500
Kapnguri	200	Leafy mustard	600	40	24,000
	250	Bean	200	50	10,000
Zothana	200	Chow chow	500	15	7500
	700	Broccoli	400	100	40,000
	600	Leafy mustard	1500	40	60,000
Darmingliana	100	Pumpkin	200	15	3000
	250	Coriander	500	100	50,000
Vanlalthlanga	400	Leafy mustard	1000	40	40,000
	500	Broccoli	500	100	50,000
	150	Tomato	200	50	10,000
Vanzinga	500	Leafy mustard	1100	40	44,000
	250	Leafy coriander	500	100	50,000
	250	Bean	200	50	10,000
Total income	**10,950 m²**				**882,900.00**

With the implementation of the technique for 10 farmers from a total cultivation area of 10,950 m^2, a total of Rs. 882,900.00 was generated (Table 16.9). If traditional farm ponds are lined with suitable recommended lining material, water can easily be harvested for its later judicious use for irrigation. The model technique is very effective in uplifting the scope of farming scenario of the region under tough hilly terrain with dual effects of water scarcity.

16.7 SUMMARY

The proposed model in this chapter was effective under the tough hilly terrains of Aizawl District and Mizoram in NEH region of India. At a total implementation cost of Rs. 10,058, farmers in this region can easily manage their available water resources for proper irrigation of vegetable crops. The beneficiary farmers of Mizoram got a good output in the first year of installation. However, limiting constrains for using the model are requirement of prior suitable site selection, nonreadily local availability of pond lining (HDPE geomembrane), and specific irrigation materials, and prolonged period of procurement of these materials. Coupled rainwater harvesting technique, Jalkund and gravity-based micro irrigation and sprinkler were effective and useful to the farming communities of Aizawl District, Mizoram, NEH Region in India.

ACKNOWLEDGMENT

Authors are grateful to National Bank for Agriculture and Rural Development (NABARD), Mizoram Regional Office for sponsoring the project; Rakesh Srivastava (General Manager of NABARD office), D. K. Mishra, K. P. Chaudhary (Senior Scientist and Head, KVK Aizawl, C.V.Sc. and A. H., CAU, Selesih, Mizoram); Project Monitoring Committee for imparting support in the process of implementing the project; and participating farmers.

KEYWORDS

- **hilly regions**
- **roof-top rain water harvesting**
- **water harvesting**

REFERENCES

1. BIS. *Design and Construction of Plastic Lined Farm Ponds: Code of Practice*; Bureau of Indian Standards (BIS): New Delhi, India; IS 15828: 2009, 2009; pp 1–8.

2. Kumar, A.; Kumar, M.; Singh, A.; Kumar, S. R. K.; Srivastava, N. A. K. *Application of Plastics in Agriculture for NW Himalaya: Water Harvesting and Utilization*; Technical Bulletin. 28(3/2007); VPKAS: Almora. Uttarakhand, 2007; pp 1–34.

3. Kumar, A.; Singh, R. *Plastic Lining for Water Storage Structure*; Technical Bulletin No. 50; ICAR – Directorate of Water Mgmt: New Delhi, 2010; pp 1–34.

4. Lairenjam, C.; Singh, R. K.; Huidrom, S.; Nongthombam J. Economic Analysis of Integrated Farming System Using *jalkund* and Micro irrigation System to Harvest Rainwater in North East India. *Int. J. Agri. Ext.* **2014,** *2* (3), 219–226.

5. Mahanta. C. *Water Resources of the Northeast: State of the Knowledge Base*; Natural Resources, Water and Environment Nexus 2; IIT: Guwahati, 2006; pp 1–22.

6. Nongthombam, J.; Kumar, S.; Lairenjam, C.; Senzeba, K. T. An Introduction to Gravity Irrigation System: Its Components and Application in Tough Hilly Terrain of Aizawl District, Mizoram in NEH Region of India. *CAU Farm Mag. (Central Agriculture University, India)* **2015,** *5* (3), 31–34.

7. NRCS and NHCP. *Pond Sealing or Lining-flexible Membrane*; Code: 521A, Conservation Practice Standards; Natural Resource Conservation Service, US Government: Washington, DC, 2011; pp 1–5.

8. Roy, S. S.; Ansari, M. A.; Sharma, S. K.; Prakash N. *An Overview of Integrated Farming System. Integrated Farming System: An Approach Towards Livelihood Security and Natural Resource Conservation*; ICAR Research Complex for NEH Region, Manipur Centre: Imphal, Manipur, 2015; pp 1–18.

9. Sharma, K. D.; Kumar, R.; Singh, R. D. Water Resources of India. *Curr. Soc.* **2005,** *89* (5), 794–811.

10. Singh, A. K.; Rawat, S. S. *Techno-Economic Feasibility of Roof Top Rain Water Harvesting System for Poly-House Vegetables in Hills*; Agri. Tech. India Bulletin 7: New Age Protected Cultivation; BIEC: Bangalore, India, 2015; pp 16–20.

11. Suresh, R. *Soil and Water Conservation Engineering*; Standard Publishers Distributors: New Delhi, 2013; pp 1086–1089.

12. Singh S. L.; Sahoo U. K., Soil Carbon Sequestration in Home Gardens of Different Age and Size in Aizawl District of Mizoram, Northeast India. *NeBIO: Int. J. Environ. Biodiv.* **2016,** *6* (3), 12–17.

13. State Meteorological Centre, Directorate of Science & Technology, Government of Mizoram. *Meteorological Data of Mizoram*; 2017; pp 7–14.

CHAPTER 17

DESIGN OF INDIGENOUS GRAVITY DRIP IRRIGATION SYSTEM FOR EFFICIENT UTILIZATION OF HARVESTED WATER

DHIRAJ KHALKHO, R. K. NAIK, S. K. PATIL, YATNESH BISEN, and D. S. THAKUR

ABSTRACT

Bastar plateau receives an annual rainfall of 1400–1600 mm. The topography of this region plays vital role in creating hindrances for suitable harvesting and water conservation from runoff. Due to undulating topography, gravity-operated drip irrigation system was designed for irrigating small patch of *plots* using indigenous materials. The technology developed in this study was a blessing for the tribal farmers with low farm holdings for increasing the production and cropping intensity by increasing the irrigated area. The developed gravitational irrigation system was able to provide adequate waterhead due to an overhead tank. During 2008–2010, tomato and bottle gourd were successfully grown with mean yield of 32.9 and 12.4 tons/ha and mean BC ratio of 11.91 and 5.01, respectively. Presently, the system is being successfully operated at 15 farm families at Bhataguda and Turenar villages of Jagdalpur block in District Bastar.

17.1 INTRODUCTION

Bastar (the tribal rich district of Chhattisgarh state) is famous for its natural vegetation and topography. The district consists of a major portion of tribal inhabitants, who were involved in faulty method of cultivation practices due to their ignorance and were confined to only a mono-cropping system.

Although the region receives an average annual rainfall of 1213.6 mm, yet most of the rainwater is wasted in the form of runoff that often attains erosive velocity due to highly undulating topography of the region. It causes soil loss, which has a very devastating effect on the fertility of the field and crop yield. Even because of ample quantity of rainfall, yet mono-cropping system is practiced in the region. Conservation structures play vital role in improving the soil condition and maintaining the fertility of the field on sustainable basis.

Suitable irrigation system for homestead garden (locally known as *badi*) needs to be designed for these tribal farmers with small farm holding. The topography of the region plays a vital role in creating hindrances for suitable harvesting and conservation of runoff. The vegetable cultivation in the Bastar plateau is restricted in the *badies* (kitchen garden) and near the *nala* (small irrigation tributary) and river banks due to unavailability of irrigation facility and topography of the cultivable land. Farmers adopt traditional methods of cultivation with little or no use of fertilizers and plant protection measures. Only 3% of cultivated area is under traditional irrigation system. Therefore, mono-cropping "rice-fallow" is prevalent. In the Bastar region, small dug wells are used as source of water for daily needs throughout the year in selected places.

Rosegrant[5] suggested a number of water management strategies in the future, such as careful exploitation of new sources of water and measures to enhance water-use efficiency. Shallow dug well and small storage tanks have been excavated under various developmental activities for land and water resources. Mere construction of water harvesting structures will not fulfill the objective of adoption of double cropping under irrigation. Proper adoption of low-cost, indigenous irrigation systems with high water-use efficiency, and less consumptive use must be designed for efficient utilization of harvested water and increasing cropped area under irrigation.[1]

Vegetable cultivation can be promoted in the *badies* of tribal farmers for improving their livelihood and socioeconomic status. One of the major issues in irrigation has been lack of power source for pumps or high initial cost for adoption of high technology micro irrigation. Due to these social and economic barriers, tribals of the Bastar region are still forced to stick to rain-fed farming only, which is an obstacle to the possibilities of the nutritional and economical safety.

The present study makes an attempt to design and fabricate low-cost, indigenous, low power requirement with high irrigation efficiency and low water consumption of an irrigation system, which can be used throughout the year for cultivation of vegetable crops after rain-fed paddy. Advanced micro irrigation system based on storage tanks was designed for irrigating 0.1–0.2 ha of field.

17.2   MATERIALS AND METHODS

Under All India Coordinated Research Project (AICRP) for Dryland Agriculture, Shaheed Gundadhur College of Agriculture and Research Station, Kumhrawand, Jagdalpur, several interventions were planned and executed for efficient recycling and reuse of harvested water, which is practically feasible and adoptable by tribal famers under rain-fed situations.

Long-term rainfall data from Agro Meteorological Observatory of the Research Station were tabulated, studied, and analyzed to determine the actual water availability for the region.[3] Figure 17.1 shows the long-term rainfall pattern for the Bastar region during the past 30 years. Table 17.1 shows the normal and annual rainfall with coefficient of variation (CV) of 14.13% during 2010.

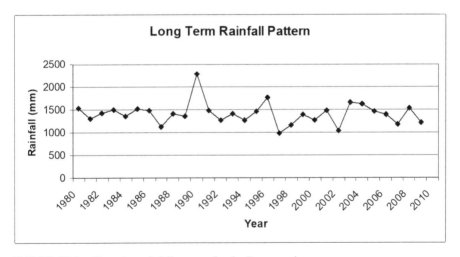

FIGURE 17.1   Changing rainfall pattern for the Bastar region.

The drip irrigation system consisted of two drums of 200 L with fitted brass tap, main flexible plastic pipe of 12.5 mm diameter, submain flexible plastic pipe of 6.5 mm diameter, medical syringes as drippers (18 and 22 syringes to maintain uniformity coefficient), T-type plastic connectors and plastic reducers. All materials were locally available in the open market. The size of medical syringes was different to maintain uniform water pressure till the end of submains.[4]

TABLE 17.1 Annual Rainfall, Seasonal Rainfall, Annual Rainy Days, and Seasonal Rainy Days in Bastar District of Chhattisgarh, India.

Station name	Normal annual rainfall	Average annual rainfall	CV	Seasonal rainfall	Annual rainy days	Seasonal rainy days
	mm	mm	%	mm	days	Days
Bastar	1413.2	1213.6	14.13	1006	87	71

Tomato was cultivated during *rabi* season followed by cultivation of bottle gourd during summer. The water from the dug well was lifted manually to fill the two drums (each 200 L capacity) daily in the evening and irrigation was allowed for about half an hour. Each plant received the water through the drippers. Daily observations were recorded for each plant.

Advanced micro irrigation gravity system was also designed with one 500 L overhead tank (500 L capacity) that was located at a height of 1.8 m for creating the desired operating pressure in the case of pressure compensating drip irrigation system. The system was designed to run at low power consumption because the farmer filled the overhead tank only once at 2 days interval.[2] As the water availability in the case of shallow dug well (size 9 m depth and 2 m diameter) has low-runoff harvesting capacity, therefore the system was designed to exploit only 500 L (0.5 cum) of water for irrigation of vegetable crops cultivated in the *badi*. It was observed that the power consumption is lower compared with traditional practice of irrigating the field by pump, which was operated continuously throughout the irrigating duration. Tomato, eggplant, and chilli were grown in the field at villages Bhataguda and Turenar in block Jagdalpur of District Bastar. This technology will be extended to other blocks for increasing the irrigated area of the region and thereby improving the socioeconomic status of the tribal farmers.

17.3   RESULT AND DISCUSSIONS

17.3.1   WATER AVAILABILITY

The water availability from the shallow R.C.C. ring dug well was observed to be limited because of low water holding and recharging capacity of the upland geology of the region. The shallow dug well helps to recharge the groundwater table of the upland during the rainy season. About 35 m^3 of rainwater was retained in the open well and is available for reuse for irrigation. If it is judiciously used, the available water is enough to irrigate 0.1–0.2 ha of land under micro irrigation. The open R.C.C. shallow dug well plays key role in increasing the water resources at the doorstep of the tribal farmer and thereby encouraging and motivating them to grow vegetable crops in the *badies* for nutritional and economic security.

17.3.2 DRIP IRRIGATION

The indigenous drip irrigation system was found to be operating at 0.64 L/h of discharge. The 22 needles (small size) as drippers were used in the upper portion of the field and 18 needles (bigger size) were used in the downstream portion of the field. This difference in the needle size was allowed to manage the waterhead at the lower portion of the field. The uniformity coefficient of the system was 83%, which can be considered good for this low-cost drip irrigation system. The experimental plot was 300 m^2 with 50 drippers on the lateral at an interval of 60 cm and each lateral has 25 needles as drippers. For 1250 plants of tomato, 400 L of water/day was sufficient compared with 1500–1600 L/day in traditional drip irrigation system. This caused 73% saving of harvested water.

The low-cost indigenous developed drip irrigation system was successfully tested at the farmer's field for cultivation of tomato, eggplant, chilli, cauliflower on 0.1–0.2 ha. The 500-L overhead tank was located at height of 1.8 m for adequate waterhead and maintaining the uniformity coefficient to >85% (Fig. 17.2). The farmer only filled the overhead tank once a day and he started the irrigation by opening the gate valve. The system saves irrigation water and labor thereby reducing the initial cost and increasing the crop yield. The fallow leftover field in the absence of irrigation facility can be brought under additional cropped area and thus increasing the cropping intensity.

17.3.2.1 CROP YIELD AND ECONOMIC RETURN

In this study, significantly higher yield of 32,945.5 kg of tomato/ha with BC ratio of 11.91 was obtained under midland and lowland farming situations during *rabi* season. Bottle gourd yield was 12,417.85 kg/ha with BC ratio of 5.06 in the summer season during 2008–2010.

FIGURE 17.2 Experimental layout of indigenous drip irrigation system with pictorial view of harvesting of tomato and bottle gourd.

17.4 SUMMARY

Proper management of harvested water though in small quantity can lead to better crop production and yield, which can help to improve the socioeconomic status of the tribal inhabitants. The indigenous low-cost irrigation method with high application efficiency was designed and fabricated for better adoption among the small farmers.

KEYWORDS

- **distribution efficiency**
- **harvested water**
- **indigenous drip system**
- **micro irrigation**
- **recycling**

REFERENCES

1. Brown, E. P.; Nooter, R. *Successful Small-scale Irrigation in the Sahel—Africa.* Technical Report WTP 171; World Bank, Washington, DC, 1992; p 79; http://documents. worldbank.org/curated/en/300091468740128116/Successful-small-scale-irrigation-in-the-Sahel (accessed Nov 30, 2019).
2. Capra, A.; Scicolone, B. Water Quality and Distribution Uniformity in Drip/Trickle Irrigation Systems. *J. Agric. Eng. Res.* **1998,** *70* (4), 355–365.
3. Frederick, K. D. *Balancing Water Demands with Supplies: The Role of Management in a World of Increasing Scarcity.* Technical Report WTP 189; World Bank, Washington, DC, 1993; p 126; https://ideas.repec.org/p/fth/wobate/189.html (accessed Nov 30, 2019).
4. Goyal, Megh R., Ed.. *Management of Drip/Trickle or Micro Irrigation*; Apple Academic Press Inc., Oakville, ON, 2013; p 412.
5. Rosegrant, M. W. *Water Resources in the Twenty-First Century: Challenges and Implications for Action.* 2020 Vision Discussion Paper No. 20; International Food Policy Research Institute (IFPRI), 1997; p 27; http://www.ifpri.org/publication/water-resources-twenty-first-century (accessed Nov 30, 2019).

INDEX